DAYS IN MY GARDEN

EVENING SUNLIGHT ON BLUEBELL WOOD

Days in My Garden

BY

ERNEST BALLARD

WITH 131 ILLUSTRATIONS FROM
PHOTOGRAPHS BY THE AUTHOR

CAMBRIDGE
AT THE UNIVERSITY PRESS
1919

CAMBRIDGE
UNIVERSITY PRESS

University Printing House, Cambridge CB2 8BS, United Kingdom

Published in the United States of America by Cambridge University Press, New York

Cambridge University Press is part of the University of Cambridge.

It furthers the University's mission by disseminating knowledge in the pursuit of education, learning and research at the highest international levels of excellence.

www.cambridge.org
Information on this title: www.cambridge.org/9781107674493

© Cambridge University Press 1919

First published 1919
First paperback edition 2014

A catalogue record for this publication is available from the British Library

ISBN 978-1-107-67449-3 Paperback

TO

MY WIFE

THIS, IN ACKNOWLEDGMENT OF A DEBT
WHICH CAN NEVER BE PAID, AND
TO COMMEMORATE
'DAYS IN *OUR* GARDEN'

Earth's crammed with heaven,
And every common bush afire with God.

E. B. BROWNING.

FOREWORD

In the following rambling pages I would desire to lead my reader with me, and dip for a space into those 'days in my garden' when the 'things' which are there shall speak.

Humanity is always interesting, but human nature alas! in all of us is frail and sadness too often comes —disappointing too, save in those few instances where great love gives and forgives. The 'things' of Nature to some are uninteresting, but when understood even imperfectly they become absorbing, and if in the pursuit of Nature's riches there is a risk of becoming selfish, her greater influence is for the best. For the billion million mysteries that she possesses are but an ever-increasing source of wonder; and wonder is the basis and foundation of worship. None can contemplate the perfection of her work and infinite and never-failing beauty without being uplifted; surely it must hallow our conception of the Designer, and bring us back to the Giver of All.

Nature in her Divine purity is the image of Deity.

Happiness, if alone dependent on those around

us, is ever charged with pain, but if we have a love for Nature and for the secrets she possesses, we have an inward joy, of which none can rob us. The more I turn to her, the more I realise the mystery of her greatness; the more I love her, the more love she returns to me;—she never fails: and in the largeness of her heart there is nothing to cause pain, but, the more she has to tell me, the less able am I to interpret her dumb utterances, or pass on the pure joy she gives so freely.

The illustrations, just taken from everyday country-side scenes (and not selected from well-known beauty-spots), will I hope help my crude and disjointed words; I realise how, like those words, they fail to convey the loveliness of the objects as I looked upon them. The beauty, greatly diminished in the photograph, suffers a further reduction at the hands of the block-maker and printer, however great their skill; yet if the combined effort conveys to the reader only some slight pleasure, and a greater power of appreciation of this exquisite world, I shall be more than satisfied.

E. B.

May 1919.

CONTENTS

CONTENTS

CONTENTS

ILLUSTRATIONS

ILLUSTRATIONS

ILLUSTRATIONS

xiv

ILLUSTRATIONS

ILLUSTRATIONS

DAYS IN MY GARDEN

Somewhere I have read of a traveller who, when ascending one of the more lonely and lofty parts of the Rocky Mountains, encountered an aged priest toiling upwards.

Astonished to find one so old in such wild surroundings, and curious to know why he should be found exposing himself alone to such risk and exertion, he sought an explanation.

The old priest then related how he had recently recovered from a long and very serious illness, during which he had experienced a strange and vivid dream. Believing himself to be dead, he had left the world and passed to the very gates of heaven, and there was met by one who to his great surprise addressed him in these words : ' That is a very beautiful world which you have just left.' For the first time in his long life he realised that he had never appreciated its beauty, so anxious and intent had he been to fit and prepare himself for the future life that he had shut out and forgotten the wondrous beauty in which he lived, neglectful and blind to all its magnificence and marvels. Unexpectedly recovering, he determined to spend the few remaining years of his life in seeing something of the beauties of the world in which God had placed him. And I fancy I see something of the look in the keen old eyes, under their long shaggy brows, as he leaned on his stick and viewed from that high elevation the spreading panorama below—he had found a new world, a new food to satisfy his once weary mind and tired eyes. I catch, too, something in his figure of disappointment at the thought of much lost, and of the resolve to see more while there is yet time, as he turns and sets his face up the mountain steep, his long garb encumbering him as he walks, yet so much a part of him that we would not wish it discarded for any more convenient, and certainly less picturesque, raiment.

And then my memory goes back to childhood spent in a lovely garden—half wild, half wilderness—where the rough and broken surface had been transformed by the hand of one who had the rare gift of future sight in planting. It was a summer afternoon and a friend had called to see my father. I had walked with them up a winding path, and, as the setting evening sunlight fell on the many fine trees, my father's companion repeatedly exclaimed at their beauty.

In one particular opening—I see it now—the slanting rays lit up a group of foliage of varying hues compelling admiration. 'Yes,' said my father, 'I can imagine nothing more beautiful in heaven than there is in this world, if only it is not marred by the hand of man.'

The words sank into my life. In my childish mind I had pictured a different heaven; but of a truth our minds cannot conceive anything more beautiful than the sublimely perfect work of God's creation: without the blot of man, for he alone it is that mars. But a merciful Nature is never tired of trying to cover up and hide man's disfiguring handiwork; as he scores her sides and scars her face, she smiles and dumbly sets to work to obliterate and even to grace his clumsy meddling. She will coil her trailing growths of creeping vine and clothe his ugly walls and angles, break down and softly bend his rigid lines, round and smooth his edges, crumble his harshness, colour and tone his crudeness. She will cast her mantle around his naked creations and clothe them with the simple beauty of age.

FIELD GATE POST, UGLY IN ITSELF, MADE BEAUTIFUL BY LICHEN AND IVY

Forces unbidden, a hundred thousand forms of life, spring as it were spontaneously to the task. The balmy breeze, life-giving sun, refreshing shower, the silent frost:—these mighty weapons of boundless power hold her secrets and work her will. From microscopic moss and lichen to majestic forest, with vegetation of every conceivable character and gracefulness, with infinite grades of form and tones of colour, harmonising, mellowing, chastening—to one end—the creation of beauty everywhere, her supreme and divinely-appointed task.

THE UNIVERSAL LAW OF BEAUTY

And so it seems a marvellous revelation of loving forethought that has devised this natural law of beauty, ever able to blot out all man's errors, and to

4

surround him with all that delights and uplifts his soul, in which there always lies, deeply buried perchance, yet still there, the inborn desire and appreciation for all that is beautiful; the created craving is there and the means provided to satisfy it abundantly.

But only to a favoured few comes the chance of seeing more than a tiny portion of the world. Few eyes have feasted on primeval forest or viewed its virgin soil, compared with those who work amid smoke and grime, and live the life of toiling gregarious man; and, when the rare sight comes, necessity often prompts him to despoil.

Still, for those of us whose anchorage is limited, there exists a world of infinite wonder and boundless beauty, the inexhaustible charm of colour on every hand, limitless life with all its marvels, the web of design; the workshop of the Mind that is at the back of all.

But where is 'my garden,' a plot perhaps think you, maybe an acre—two—or three? Shall it be

'MY GARDEN' WHERE IT IS

> Where lines are drawn 'twixt flowers gay and weeds,
> Concealing hedge, with paths, and beds of seeds,
> In order prim and neatness all must grow,
> The trees in pairs and plants all in a row?

5

.....FROM WHERE THE DAISIES AT OUR FEET STRETCH OUT TO SUNLIT HILLS, AND ON TO WHERE THE LIGHTS OF PALEST BLUE HORIZON MEET THE SPAN OF HEAVEN AND MELT BEYOND IN BRIGHTNESS

No! it is not alone to our own little hedged plot—
that world where we are free to do just as we like,
where every blade is ours, where much happiness
can be grown, and many sorrows buried, that I
would take my 'gentle reader' (an old expression but
a happy one). I would lead him further on, and into
my garden which knows no hedge nor bounds, no
fence nor wall, away over the wooded fields and along
the lanes, the rolling downs and sunlit hills, to where
happy, babbling brook with mossy bank and poly-
podded stub, joins the flowing river and ripples on—
a silver gleam—to open sea of liquid blue ; nay even
further, and on yon distant shore drink deeply of the
beauties there displayed.

For while we have eyes, none can deny us the joy
of all earth's coloured charms, and we can roam from
where the daisies at our feet stretch out to sunlit hills,
and on to where the lights of palest blue horizon meet
the span of heaven and melt beyond in brightness.

Come then, let us wander down the leafy ways and
smell the new-mown hay, pluck the frail and fragrant
dog-rose or trail of yellow woodbine. Let us wander
along the hazel lane, across the murmuring brook,

THE MURMURING BROOK

where the startled dipper checks his rapid flight and, alighting on some slippery pebble, elegantly bobs at our intrusion. Clothed in his almost black tail-coat and white dress shirt, he always looks a perfect little gentleman; but alas! I fear he is not. Then pause awhile and drink the sweetness of great scented fields of beans, whose silvered green would seem to have captured the moonlight sheen and to have absorbed its lustrous rays into their substance. Or listen to those old familiar sounds of summer nights, the whirring burr of nightjar, ebbing, flowing; the harsh metallic scratch of corncrake, now here, now there—quaint ventriloquist is he—while yonder whistling lad lets swing the oaken gate, and by its muffled wooden note has startled up a screaming peewit to wing its weird night flight; we hear the swoop of wing and rush of air, that strange repeating, half-grating noise, as if its wheeling wings had rusty joints; a hundred dear familiar sounds, the music of the night.

Nay, what is Nature's
Self, but an endless
Strife towards music,
Euphony, rhyme?

9

DAYS IN MY GARDEN

Trees in their blooming,
Tides in their flowing,
Stars in their circling,
Tremble with song.

God on His throne is
Eldest of poets :
Unto His measures
Moveth the Whole.

WILLIAM WATSON.

The problems of origin and destiny, of life, mind, and matter, press closer on the thoughts of each succeeding generation ; but the riddle of beauty, its source and end, is to us at times as deep and wonderful as any. Beauty exercises such a sway over us, and is so widely spread upon earth and sky and water, that we must be in a dark mood to question some highness in its origin and end.

G. A. B. DEWAR.

THE SHORTEST DAY

THERE are some days about which much has been written and justly so. St Valentine's day, May-day, Midsummer-day—they are times when Nature just comes and takes our heart and makes it larger—larger by filling it with joy—yet there is another day, often forgotten, which should be called Hope-day, or the

10

Day of Promise. For as the greatest darkness often heralds the light, so December's shortest day ushers in the Day of the Dawn.

What matters the driving rain and dismal fog or blizzard's blast and biting cold? Vanquished their power, their hour has struck, the tide has turned and every tick by night or day brings nearer the onrushing victorious sun. Higher he mounts, and, though as yet the sleep of Nature prevails, her slumber is broken, for she has heard the call go forth to every dormant bud and resting root to mobilise. Tardy perchance her response may be, still the silent push of growth will lose no chance—and buttercups of burnished gold are coming.

THE PUSH OF GROWTH

AND the insistence of this push of growth is very wonderful; it is not always that

> When the days begin to lengthen,
> Then the cold begins to strengthen

and, should the younger days of the year prove mild and sunny, it is visible in many places: not so much seen in the open country, as in the woods where a favourable aspect and the brushwood undergrowth afford encouragement and protection.

DOG'S MERCURY

There amongst the disappearing leaves the dog's mercury is in rapid growth thrusting up its fresh green heads, cleverly bent in loops of strength, soon to straighten out and unfurl leaves of tender green, revealing tassels of tiny angular buds,

12

which only need a hint of sun to burst and show their bunches of yellow stamens.

Though frail, it breasts the sudden spell of frost and grasps a longer span of life than its companions by sleeping less; for autumn finds it still at work while they who later came have done their work and gone. Others too are wide a-wake, and under shady banks the sturdy tufts of wild arum leaves, five inches high, are fresh with shining green, but best of all are primrose roots whose pointed crumpled leaves hold richest promise, and here and there, perhaps, a short stalked blossom. On yonder bush the trails of woodbine are flecked with leaves washed in silver sheen, while all around the swelling buds, now richly hued, proclaim the flowing sap at work,

ALDER
CATKINS

SALLOW BUDS

13

until the check of frost shall intervene. Down in the swamp the alder catkins grow dusky red, and sallow buds have burst their rich brown masks and show their downy smoke-white points of silk, while many a nut-tree bough holds shaking tails of yellow cord, pale as the frosty moon, to waft their golden dust on waiting buds, all carmine starred—and yet it is winter.

LAMB'S-TAILS

14

TASTE

THERE is no accounting for taste. A heavy fall of snow followed by a thaw has made most things and most people miserable. A strangely weird piece of humanity I found cutting down or 'laying' an old overgrown neglected hedge; as the blows of his 'hacker' fell on the stems of the overhanging boughs, wet melting snow shattered upon his head, his old bent back and down his neck; his scratched and bleeding hands were wet and cold, his feet deeply buried in the sloppy snow. Not far off a number of school boys were tobogganing with an ecstasy of delight and much noise. He paused—'That job seems to go well' he slowly remarked, indicating the boys, 'but I would sooner have mine,' and, turning his stiff old body, far less able to bend than the hedge stakes, he continued pursuing his operation with surgeon-like precision. No—there certainly is no accounting for taste.

15

THE BEAUTY OF WINTER DAYS

GREAT TITS
FEEDING ON
COCOANUT

The bird on the right which was feeding first, has his mouth wide open hissing at the arrival of the other. These quarrelsome habits belong to winter days and disappear with the approach of spring.

WHEN the magic work of spring has begun, one often hears the remark made, 'The country is beginning to look nice now!' Some people are very benevolent, and the way they patronise Nature and commend her efforts is very kind. Poor old Dame Nature! She is too old to blush, but she is so kind that she must be sorrowful.

How often have we seen a town, even on a spring or summer day, when we have had but one desire—to get out of it—but I have never looked upon the countryside even on the worst of winter days and not seen much of wonder and of beauty.

16

THE BEAUTY OF WINTER DAYS

The indescribable charm of spring flowers; to bask in summer sun; the golden glories of the autumn tints : inhuman, lost, the heart that does not respond to these. Yet many eyes only see the outside of those doors which open wide to lands and worlds where beauty reigns supreme and mysteries invite, where perfect laws reveal the heart of Nature as she woos and beckons along those roads which have no end, whereon the traveller never tires, where she flings wide open her caskets filled with gems and strews them at our feet, and feasts our eyes, enchants our ears, melts our hearts, and makes us know the finger-prints of God.

Gone now the milk-white fog, its vapour here is free from sulphurous soot and grime; no trace of haze to blur the atmosphere; it is on such days that winter's sunshine gives us lights and colours as beautiful as any in the year.

The upward path follows the track of one of those old and long disused roads, now but a burden scar; time has rounded off its sides and filled its ruts and ditches, and we tread a smooth grass-way.

We try to picture the tired feet of man and

B. — 2 17

WINTER SUN ON GREY BOLED OAKS AND RUDDY STEMS OF PINE

beast, who wore its hollowed groove to dust; gone
their life, their ways, their fears and hopes, but down

A BURDEN SCAR

the years have come the heritage of many labours, of
battles fought and won; their victories are ours, to
guard and keep for others yet to be. And so I love

the oaks that skirt its curve, for they have never changed their ways, have still the same strong look as they had then; they heed not war nor fashion's vogue, but fulfil their span of life to serve and beautify.

Anon the top is reached and we look down to where below a broken earth is seen, a tangled mass of vales and steeps, all pitched and furrowed, twisted deeply in chaotic maze. It is here as if the great Creating Hand had taken up

BROKEN WOODLAND COUNTRY

THE TOP OF THE BURDEN SCAR

the crust of Earth and crushed it to a ball, as we would take a silken handkerchief and, then releasing it, watch its crumpled folds relax. Then, growth had come and clothed the naked scene; now, every canted face reflects a different hue: no mask of summer

20

green, for winter's coat is brown and grey, and fields and woods are toned in every shade; a far more varied view. There where the cloud-shadow falls on distant wooded slope, 'tis purple, rich and rare, while nearer banks are sheened in deepest blue ; there are ultra-marine patches, blues and purple-browns, bright rusty reds and copper tints, the greys of boles of oak and beech, the silver-white of birch where fawn-coloured knolls crop out to drop away in sudden fault, while many a twisted fallen tree, green ivy-clad, clings to the slipping bank above the tiny brooks that hasten down the hollows.

And as the days speed on and winter slowly dies, activity of tree life quickens and every day transforms its colours and in rapid se-quence buds and barks are painted all anew. The silvery white silk points on the goat

THE SILVER-WHITE OF BIRCH

LENGTHENING
DAYS

21

willow have become exquisite soft grey pussy buds, round cushions hastening fast to grow large enough to carry their array of gold-headed pins.

Foremost among the nobler trees the common elm now leads, while oak and ash and beech seem hardly yet awake. Far above our heads in the great domed tops of the giant elms there is a thickening of the highest tiny twigs; they are studded with little round bosses of flowers, each fringed with a mass of almost crimson anthers, the colour of their myriad hosts unseen unless looked down upon, except at times when their beauty is revealed, sunlit, against a background of snow-laden clouds all leaden grey: just one of Nature's many colour schemes.

PUSSY BUDS OF
GOAT WILLOW

ELM IN FLOWER

22

IN THE TOPS OF THE GIANT ELMS THERE IS A THICKENING OF THE TWIGS

Upon the heath the aged hawthorn bush has buds, mere carmine specks, amid its lichened boughs. Sweet-chestnut tips are crimson-brown, the barks of lime and birch boughs richly-hued, the pointed cones of the larch belt golden in the sun, while just below in welcome contrast is the sombre yew, and the osier bed all yellow as the gorse in bloom or, sometimes, flaming red.

When the rabbit and the squirrel don their winter coats, sleek and warm, and the bullfinch and the robin's breast are bright and ruddy, there is scarce a colour known to art that does not deck the cock pheasant in his metallic brilliance and lustre; and the sheltering woods and trees—his home—cannot escape Nature's brush, which paints their plumage too, in colours chaste and finest hues, for all must wear a rainbow robe to welcome in the spring.

RED LETTER DAYS

CYCLAMEN

THE FRINGE OF THE DESERT

AMONG those days which I do not forget and
which when remembered always act as an anodyne,
is one when, standing on a shoulder of rocks which
jutted out from wooded hills, we looked down on
a brilliant sunny day in February to the rich green
shores of Northern Africa where they meet the
waters of the Mediterranean; the blue sky dipped

25

into the more varied and richer blues of the sea that
bathes the land where lifts

The fronded palm in air,

DATE PALM OASIS

while here and there the dazzling whiteness of a
rounded dome or Moorish arch shone out amidst the
vegetation.

26

RED LETTER DAYS

The night before, the dinner table at the hotel in Tunis had been decorated with the daintiest of wild cyclamen; and, resolved to seek out their hiding place, an early train had landed us at the foot of pine-clad hills, whose broken surface looked promising. Surely here their home would be, and as we wandered up the shady slope, noting many a pushing growth and foliage of unfamiliar form, it was not long before we espied the sought-for heart-shaped leaf among the stones. A little higher up and the vision burst upon us; yes, there they were in full-flowered wild profusion; the mountain side was carpeted with the marbled leaves with their stalks and undersides all reddish-purple.

CYCLAMEN

I know of few things in life which give us such moments of delight as is experienced when we look for the first time upon a mass of any flower growing in wild profusion, or when we find perhaps only a single spike or so of some long-sought rarity; in those few moments life overflows with joy—just because we worship. Such 'splendid occasions' are 'beyond words to express, for one ecstatic moment we seem carried beyond the mundane plane

27

of self-consciousness and launched in the realm of Reality.'

Think of the exquisite delight of gathering great bunches of those chaste tiny gems, each pure white twisted petal with its base tipped with brightest crimson, some more, some less, while in a few the rich colour had spread and suffused the whole flower with a dainty pink; these, and an end-less variety in the lovely marbled markings of the foliage, made us careful in the selection of plump corms to fill our vasculum, never large enough, but never yet so

CYCLAMEN AFRICANUM

full that it would not hold just another, more longed-for than the last; they could not be missed from among those millions and when home they would not lack love: if they would only live. Here they flourish right under the trees, thriving in the almost black mould

28

that had accumulated between the loose flat stones, growing in many a quaint position beside the mountain stream, where small avalanches had buried some so deeply that they had thrown up a long wiry stem, and then produced their leaves and flowers. Here, too, were hosts of tiny seedlings, with pale, almost transparent, little globes, shell pink, with baby leaf and stalk.

As we wandered on, a bend in the upward track brought us suddenly in sight of a small company of Monks—'the little white fathers.' They will always be associated in our minds with *Cyclamen africanum*, for as they came down the mountain path clad in white serge, crimson fezes on their heads, they seemed in wearing these colours to have adopted them from the dainty little flower which, decked in almost identical tones, surrounded their mountain home. Alas! this lovely African gem is too tender for our gardens, but its cousin *neapolitanum* is so hardy and so easy to grow that none should miss it. Accommodating itself to sunless positions, around the boles of large deciduous trees, even under the dreaded 'drip,' in a little leaf mould, old mortar and stones, it will

29

quickly make a home and is worth growing if only
for its exquisite and varied foliage in winter
and spring; but when September comes it
produces the most charming and delightful
profusion of fine pink and, more rarely,
pure white flowers.

CYCLAMEN
NEAPOLITANUM
(hederæfolium)

CYCLAMEN NEAPOLITANUM GROWING AGAINST THE BOLE OF
LARGE ELM TREE

FIRST SPRING FLOWERS

. . . . . . earth is a wintry clod :
But spring-wind, like a dancing psaltress, passes
Over its breast to waken it, rare verdure
Buds tenderly upon rough banks, between
The withered tree-roots and the cracks of frost,
Like a smile striving with a wrinkled face.

' Lo, the winter is past, the rain is over and gone ; the flowers appear on the earth ; the time of the singing of birds is come.' So runs the song of the Wise Man, and if we are too prone to harbour the delightful thought that winter is past and a new season at hand, we may be forgiven, for a mild and visibly lengthened day has followed one of blustering wind and icy snow and we have found the first golden globes of the aconite nestling in their green ruffles : and in its tender elegance have found a snowdrop ' blanched

THE FIRST GOLD GIFT HAS COME TO BLESS THE YEAR FROM EARTH'S ABUNDANT BOSOM..........'

'BLANCHED WITH FRIGHT'

31

with fright' that has outgrown the palisade of little spiked leaves around it, defying frost and drift of snow.

And within us has come a feeling, a movement of the roots of our garden-love, a stirring, quickening, fluttering; the fascinating touch of the damp moist soil, the crackle of the seed packet, the peculiar bursting noise of the clump of perennials as the handles of the two dividing forks meet, the sharp click of the secateurs, a hundred memories of happy sounds and proceedings are awakened and will not sleep again nor be discouraged by the intrusive thought of aching backs, sore hands and tired limbs. While the Christmas rose was with us it was winter, but when the first little knob-heads of pink and blue hepaticas, which have

HEPATICAS

FIRST CROCUS

32

managed to evade the hungry mouse, are opening, and
the first orange-gold crocus tempts the mischievous

SNOWDROPS

sparrow, why then it is spring—at any rate spring
within us—and the heart is young again.

There is a halo of charm around these first
spring flowers of the garden, a feeling of a
faithful tryst, and we stoop in ecstasy to
greet the brave forerunners who, while
winter still commands, have outstripped their
comrades and hastened to proclaim, even at the
cost of their own life's cycle, that just below the
cold bare ground lies hidden the bursting array of

ACONITES

Mother Nature's spring carpet, in familiar and yet ever new pattern. And, somehow, we find as much pleasure —nay more—in the tiny vase of these earliest heralds which we rescue from the still unkind elements as we shall find in the bowl of June roses. We watch the roses as they grow, we see them slowly come as sunny days grow longer ; we tend and prune and wash and kill, we watch the promised unfolding of the blooms ; not so with these first spring

IRIS RETICULATA

gems. Only yesterday, the frost-bound soil showed but the tips of green-spiked blades of *Iris reticulata* and to-day, as if some fairy's

THE SAME CLUMP TWENTY-FOUR HOURS LATER

34

wand had loitered there, we find its gorgeous blooms of gold-laced regal purple, or as if some specks of heaven's deepest blue had fallen low—the scillas.

SCILLA
SIBERICA

Dainty *Scilla siberica* has many charms and many blues, real blues, china blues that make us think of cups and plates on dark oak shelves, of low panelled rooms where everything speaks of an age when people had more time, or made time, or perhaps knew how to use time, when we half-fancy clocks went slower and hearts were truer and feelings deeper. Yet our scilla was the same, we are sure they loved him, like them he never failed, and now he brings a link of chain to hold and bind our hearts to them and him. For they had not seen the melting snowdrift fringed by gentian-blue, had never plucked the tiny soldanella; brain-fag, then unknown, needed no such sight—now February's first blue squills no longer satisfy—we have progressed!

It is however later in the boisterous April days

SOLDANELLA

35　　　3—2

that the scilla is at its best, no longer stunted, now it rises in all its pride and beauty, throws high its head all starred blue, mingling all shades of dark and light—it knows no rivalry of crews of eight— but unabashed wears both colours, yet always true blue, almost shaming its stiffer cousin the hyacinth, whose heavy head and tender juicy stem often falls a victim to April's sudden storms. At home, in half-shade, the scilla makes a blue carpet, no mean setting for the blooms above, and multiplies its race in tiny blades of grass-like green, perhaps to meet an early death by unrespecting hoe.

TREE BOLES

To me there is an ever increasing delight in the boles of trees, more especially in the massive beauty of the deciduous trees, never omitting however the handsome pine or Scotch fir. At every season they retain their attractiveness, their varying colours, the marvellous beauty of their individual barks, with corrugations and scales, often intensified by bough scars, some screwed and twisted like the pear and Spanish chestnut, others mathematically symmetrical, or charming in their irregularity. Note the contrasts of the circling light and shade on their rounded boles, the dappled patches where the filtered shafts of sunlight fall, the colours of the algae and lichens which embroider and enrich their surfaces: yes, they are superbly beautiful and wonderful.

We need but a very slight knowledge of leverage and weight, of stress and strain, to appreciate something of their gigantic strength. Try for a few moments to hold out your spade or lighter rake at arm's length and see what it means to support its in-

37

GIANT BOLES OF THE SPANISH CHESTNUT

significant weight, then gaze aloft among the mighty limbs of a giant oak, see where the monster thrusts its huge outspreading arms far out, to carry and support a thousand subdivisions of its 'kneed' branches; watch the upward swaying lift as the gentle breeze caresses and you will marvel at its colossal strength; think then what it must be to sustain the full force of the roaring gale. Look at the wondrous balance and spacing of the eyeless limbs, as their tapered shafts extend; and this is all achieved with never-failing grace and beauty. We marvel at the germ, that tiny seed, from which it sprang: no hollow beneath the spreading limbs of used-up soil is seen, built, it would almost seem, of air and dew and light and Heaven's Will. One of Nature's grandest master-pieces: thus skilfully she works combining great strength, utility and perfect beauty, a task wherein man most surely fails.

We are accustomed to connect gracefulness with a certain slowness of movement, at any rate in human beings. The graceful woman usually moves slowly, the graceful dance charms with its slow rhythm, and

THEN GAZE ALOFT AMONG THE MIGHTY LIMBS OF A GIANT OAK

the sudden or unexpected movement is seldom captivating, more often grotesque. But see the startled stag leap up the forest glade with perfect grace and ease, the poise of head and neck, the delicacy of limb doubly refined in motion, swift and strong. Or watch the almost lightning movements of the common water wagtail on the lawn, himself the daintiest combination of agility and grace, his rapid runs and natty twists; the ever dipping tail, yet no faulty turn nor ugly hop, no graceless step, every movement one of refined elegance. Compare his agile, shapely form with that of the tame pigeon, who clumsily struts and hustles along full of awkward importance (from whom could he learn such ways?); 'twould almost seem as if Nature had been unkind, but no, the unkindness rests with us for making the comparison; for life always holds its compensations, did we but see and know them.

COLTSFOOT

COLTSFOOT

ONE of the earliest of flowers and the first to make a brave attempt to heal the ugly earth scars inflicted by man, is the beautiful coltsfoot.

Inside our garden ground he is a troublesome fellow with deep roots enabling him to persist most aggravatingly, but the readiness with which he establishes himself in the uncongenial barren soil exposed in the cutting and embankment, or on the spoil-bank, is marvellous.

Man, not content with scarring the earth, often commits the further crime of burying the priceless top-soil, a heritage handed to him which he has no right to destroy nor in any way injure for future generations; never his own, he is but its tenant, it is his but to till, to enjoy and to pass on.

Planting itself in unproductive subsoil, the coltsfoot begins Nature's task of reclaiming and beautifying. The big-flaked snowstorm of February has hardly melted when the lovely fringed yellow flowers star the bank, mere specks asleep on 'fill-dyke' days,

42

but no sooner has the sun burst through than they are wide awake to welcome him, and open out their finely cut florets which, though dependent on, are seemingly neglectful of, the later crop of large and handsome leaves which they never see ; silk-backed, these clothe the naked earth and by their work begin to weave again its beauty and win it back to use.

COLTSFOOT ON THE SPOIL-BANK

OLD PETER

THERE are two words connected with the garden, they belong to a big genus and in their frequent use comprise many species. One is the word 'hardy.' How it grasps within its long cold arms a great army of the robust and strong, together with the tender, delicate and weak! How large is its embrace judged by the plant catalogue, but alas! how we mourn to find many an isolated label; a tombstone after a severe winter, or perhaps, even more frequently, after a mild and wet one. 'I am hardy anywhere;' 'almost hardy, if—' 'I was hardy before that winter;' 'I shall be hardy if planted in a protected spot;' 'perfectly hardy, note, I winter under glass': we learn these painful qualifications by experience.

The other word is 'gardener.' How many claim that distinction; how few attain! There is all the difference in the world between 'a gardener' and 'a man about the garden,' the latter is common enough. Dean Hole said 'He who would have

44

beautiful Roses in his garden must have beautiful Roses *in his heart.*'

Therein lies the difference. The 'man about the garden' plants and sows: he even obtains a certain amount of success, but it is his 'job' and not a task of love, he does it for a living, his ends are time by the clock, and wages.

The 'gardener' gardens because he loves his flowers, fruit and vegetables. They are as his children and he counts neither the time nor trouble spent upon them.

I used to think Old Peter was just 'a man about the garden,' certainly he always is about it morning, noon and night. He keeps no clock time, but it was long before I discovered that he grew two things in his heart—a strange couple too—potatoes and red currants, or as he would say, 'taters and red currans.' He is keener and more observant of the skin, shape and eye of a potato than he is of his mates, and his love for the sour and gritty little red berries is amazing. 'I have twenty-one currans on one bunch of Fay's' he excitedly announced one day; a never-to-be-forgotten achievement.

It is there that his claim as 'a gardener' comes in, this old son of the soil, but he is 'not much of a hand on the flower knot,' as he one day owned.

It had been the longest winter I had ever known and snow fell day after day in early April; planting was impossible and I began to lose heart; almost

> I dream'd there would be Spring no more,
> That Nature's ancient power was lost;

not so Old Peter, he had asserted that the wind was wrong on Candlemas Eve and therefore it was inevitable that we were in for a bad time; the wind, be it known, according to him, only blows from two quarters, 'up-hill' and 'down-hill.' It was 'up-hill,' that is north or east, on Candlemas Eve and we should have no good weather till 'Old March was out,' that is April 12th, and even then we might get 'lamb's snow'—and Old Peter was right.

The cruel wind ceased at last, it seemed to blow itself away and there were days of perfect calm. In the still air the sun attained his full midsummer power and with one mighty bound spring came, throned on a carpet of loved old, but ever new,

46

magic pattern, the woods green-washed like changing pictures on a screen, and bursting buds were leaves.

By lightning strides Nature hasted to make amends for her lateness of which she somehow seemed well aware, and now I dreamed no more, for surely I knew that 'Nature's ancient power' was far greater than I had ever thought.

ALDER CATKINS

47

RAINY DAYS

MARCH

RAINY days in 'lion March' do not have the depressing effect of soaking, songless, autumn days, though too many of them are aggravating at a time when there is much to be done. We feel the advance of spring, it cannot be stayed and, despite

GREAT-TIT

the steady downpour, the birds persistently sing their joyous notes. There is an absence of falling, drifting leaves; a neatness in the fields and a general appearance of tidiness about the trees. These last, awakening from the deep sleep of winter and bathing in Nature's

ALMOND
BLOSSOM   bath, shake their great limbs like some prehistoric mammoth in the drying winds; their strength renewed.

WITH BLOSSOMS HAVE
BEDECKÈD DAINTILY (SPENSER)

48

RAINY DAYS

The delightful sight of odd patches and specks of new pale green in the hedgerows, the splashes of many-coloured crocuses, the alertness of the daffodil leaves, the bursting buds, all ready for the forward movement: it needs but the cloud-curtain to rise

LAMBS IN AN ORCHARD

and sunlight to flood the scene, and spring will have made another bound, the 'lion' has gone the 'lamb' has come. The great-tit, 'spring's early bellman,' sounds his note where the almond buds have opened wide, and blossoms of the daintiest pink are outlined in the blue; while down the orchard bank the lambs

run wild races, their stiff and clumsy legs too large
for such gay frolics ; but youth comes only once—for
them but one short, quickly passing spring.

SEEDS

OUTDOOR gardening is impossible when the thaw
has come after frost, for the soil is a sticky mass, the
air is unpleasantly raw and we can only
garden in anticipation. So, great is the
pleasure when we take the recently
received stack of nurserymen's lists
and, perusing them, mentally order our
new plants and seeds.   Then it is that many
things grow and flourish by the fireside, perhaps
because they are warm and free from blight, for there
is no insecticide like nicotine !

It is a long list of seeds that we intend to grow
this season and the time for planting is almost upon
us.   The time of seed planting !—the miracle when
we are able

> To watch the matchless working of a power
> Which shuts within its seed the future flower.

SEEDS

Look, if your eyes are good enough to see it, at the minute dust that escapes from the bursting capsule of the orchid; unfold with care the final small wisp of paper in the inner packet of gloxinia or begonia seed and think of the superb Odontoglossum spray, the spotted long deep throat of the Gloxinia and the lovely flushed bloom of the Begonia that is 'shut within.'

What subtle power is there in these specks of brown dust, these tiny germs, which are able to extract from the moist earth a mere trace of mineral salts and, with the sun and air, build up a myriad forms of beauty, all modelled with varied grace, painted with a delicacy or richness of hue, perfumed with a sweetness, exquisitely perfect!

THE HANDSOME SPIKE OF THE BULRUSH AFTER PROTECTING ITS SEED ALL THROUGH THE FROSTS AND RAINS OF WINTER DISPERSES IT ON THE FIRST WARM DAY OF SPRING, BURSTING INTO A CLOUD OF THE MOST BEAUTIFUL WHITE PAPPUS; AS BROWNING SAYS:—

'......A MARSH
OF BULRUSH WHITENING IN THE SUN'

Whence the 'power' whose 'matchless working' takes the seed and, with these same tasteless things, produces under equal conditions the succulent pungency of the onion, the sweetness of the carrot, the delicate flavour of asparagus and pea or, with longer time, the crisp, juicy apple or luscious pear?

Where within these microscopic atoms lurks the wizard who, as the blooms unfold, distils the fragrant scent of rose, mignonette, jessamine and stock?

There is no change in the sweetness of the violets as spring after spring they come and go, the recipe is never lost, the formulae never vary, for, mark you, Nature never forgets anything. The wafted sweetness of the lime trees laden with blossom, the bloom upon the grape, the flavour of the peach, the graceful curves of the delicate pink-coloured 'Lady-in-the-Boat'; these things change not, fail not, and we are like those who marvelled at Goldsmith's village schoolmaster :—

> And still they gazed and still the wonder grew,
> That one small head could carry all he knew.

So we ask ourselves again, how is it possible that this seed dust can 'shut within' its minute brown

SEEDS

walls a whole being of life and character, the personality of the plant, its flower and fruit?

From the wee seed of the black poplar, less than a grain, springs the giant tree, a hundred feet high and weighing many, many tons.

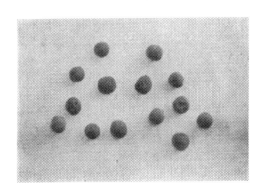

SEVEN SEEDS (Left)
SWEET-PEA

SEVEN SEEDS (Right)
COMMON VETCH (V. Sativa)

Seeds are endless in their variety of shape, colour and beauty, while their wisdom in movement, dispersion and protection, their art in deception, surpass human imagination.

Many are so small

VETCH   SWEET-PEA
EACH PLANT GROWN FROM ONE OF THE SEEDS SHOWN ABOVE

that when mixed together they cannot be separated until growth has revealed their identity, not even when as large as those of the sweet-pea and fodder vetch ; yet what a difference is seen in the fine long stalks and bold flowers of sweetly scented white, red and mauve sweet-peas, with their strong stems and leaves, when compared with the small, insignificant, stalkless, rose-purple flowers of the useful vetch, with its finely-cut leaves and hair-like tendrils !

It was summer, and the cool, clear stream threaded its way through the shimmering meadows and passed under the shade where silvery, slanting willows bathed their feet, all clothed with deep-red fibrous rootlets. In the shallows, near the bank fringed with yellow iris, there were strong colonies of watercresses with their deep glistening green and bronze foliage, while just above rose the tall furrowed red-dappled stems of the graceful hemlock, carrying their elegant feathery leaves. Again the problem comes, why should one plant produce from the same water and soil the wholesome salad, and another spend its strength in extracting a deadly poison ?

54

SEEDS

I took the upward path that led to rising ground
and, passing through the hazel copse that flanked the
limestone bank, I found the dusky purple bells
of flowering Belladonna, while from the self-same
earth the young nuts filled their soft green
shells with snow-white pith, already showing signs of tender kernel.

I climbed still higher to where the
fibrous soil lay deep upon the igneous
rock; a hillside squatter's garden plot
held within its straggling elder fence a patch
of fine potatoes and silver-green cabbage, but
from the untrimmed fence by the garden path
a hundred giant spikes of lovely foxgloves waved
in stately beauty, masking the death they held within.

How many a similar problem lies hidden locked
in Nature's casket! Broadcasting her seed, in what-
ever aspect and soil it may chance to fall, she produces
therefrom unlimited variety, for the use and need of
man, working as ever with schemes of supreme beauty,
and we seek in vain for material out of which to file
a key, whose wards will unlock her mysteries, and
reveal the inner secrets of her wonder working.

YOUNG
HAZEL NUTS

55

DAFFODILS

Some come from gardens and stand proud as princes,
And many will tell you that those are the best;
But the dearest to me are the shy ones, the wild ones,
The daffodils with short stalks that grow in the West.

Is life worth living? Is life good, has it any sense of completeness or satisfaction? There is but one answer on an early April morning, when the cold wind has gone and the sun has suddenly grown warm; winter is vanquished and the whole atmosphere is charged with spring; we feel it, see it, smell it. We hear it in the first broken notes of the piping blackbird—fragments of melody; he pauses, hesitates as if not sure of himself, half shy, as if he had just realised that in a songless winter he had forgotten his art, then, out throb the full tones, liquid in their mellowness and free from all trace of metallic twang or scratch, of which the song thrush and other songsmiths are sometimes guilty. Again he pauses and the still air is broken by the cheerful, sprightly notes of the chiff-chaff. He is our earliest and so most welcome of summer visitors, at any rate to our inland woods in the west. Yes—down in the west in April, that

56

DAFFODILS IN THE WOODS

is where the daffodils grow, where they bloom in lavish profusion, and there I have wandered. They are out in the meadows and orchards, great breadths of them, massed yellow patches, or single outposts scouting the hedgerows; the advance guards prospecting new and untried soil, preparatory to colonising. But these hardy yellow lilies of the open fields have short stalks as yet, and are later than those which fleck the neighbouring wood—there they are tall and finer, great sturdy fellows, but never coarse, never unrefined, which cannot always be said of those which figure at our spring shows. How well the natural carpet of dry brown oak leaves shows off the neat, straight tufts of spiky leaves!

Daffodils, yes, daffodils by the million million, singly, in groups, massed in battalions, each a perfect flower, regiment after regiment, the whole wood is tessellated and dappled with them. Where the young unfelled oaks are thinnest, and the undergrowth cleared they crowd and run riot: a glimpse of the well-trimmed ride shows just a lane of gold, while half-way down gleams a group of silver birch stems, almost snow-white as the sunlight strikes them. On they

58

WHERE DAFFODILS GOLD-PATCH THE MEADOW LAND

run in stippled patches, all with their trumpets turned sunwards, on and on till the numberless multitude form one solid yellow sheet. In those near by, the individual grace and beauty of each flower stand out and the contrast of deeper yellow trumpet and lighter yellow petals is individually pleasing, but as they recede from the eye the colours mingle to one shade, growing paler and paler till in the distance they are primrose-hued.

And, as my eye has wandered down the lane of gold, there has come into my thoughts the shallow and unattractive expression 'streets of gold.'

If in that other land, where there is no pain, there are streets of any kind, I think they will be of yellow daffodils, rich buttercups or even more majestic king-cups, and a thousand other hues. Here, we may for a time have to acknowledge the use and power—aye, the often cruel power—of the precious metal, but I like to picture a land without a street with its limitations and restraint, where gold and its power are gone for ever.

A DEAD BUNNY

It was just an old packing case, with part of the
open front covered with wire netting, under a wall in
a sheltered corner of the garden, but it was
the home of a wee wild bunny.

It was when the snow was deep
upon the ground that Old Peter in
moving a heap of sand had destroyed
the nest, and exposed two tiny furry
mites; too small to feed themselves, one
soon died, the other learnt to use a spoon
and so saved his life and grew.

RABBIT HUTCH

Only yesterday—Good Friday—
he was gay and lively, with his
dark soft brown eyes and little
twitching whiskered nose; as for
the pace of his rapidly moving jaws, 'twas wonderful;
no sort of weed to him could come amiss.

Alas! to-day he is dead. I know not why, but it has

made me sad to think his little happy life has ended
—yet he was only a half-grown bunny, nothing more—
to whom death had come
too soon it seemed. Wild,
doubtless a few more
months would have
brought his life to an
end, the pounce of
spaniel amongst the bushes,

a beater's shout, a loud report, and one more
bunny to add to the couples in the bag,
and none would have regretted him. But
now—now on this April morning, when
all the buds are bursting and springtime,
real spring, is everywhere—I do not like
his death; a death in spring. Why should it
snatch him thus away?—
and—not him alone
—and I am more sad
until I remember,
to-morrow is Easter-
Day!

PRIMROSES

I<small>T</small> is

> Whan that Aprille with his showres soote
> The droghte of Marche hath perced to the roote,

but his cold wind is still a-blowing.

I have wandered over a wind-swept stretch of grassy banks and fields, gradually ascending until the turf has put on its fine wiry hill-dress, very different from the young green meadow grass or the rank lush growth of low-lying land. How often we are reminded that grass is a water-loving plant, how green are the valley meadows, but how soon a short summer drought has 'burnt' the hill, and the farmer says that 'keep' is scarce! The velvet lawn is patchy and brown, the landscape becomes uninteresting, our garden is a desert and for a moment we almost forget ourselves and whisper that the burning sun is an enemy; but when the water-pots of heaven flow the change has come; as if by magic, the grass, the earth are green again.

But I have topped the low hill and, passing over, leave the wind to roar like a wild sea in the tops of the trees which form the long straggling ridge of the wood below and, as I step down its cleared and sheltered sunny southern slope, how can I describe or give an impression of 'my garden' here? I have dropped into primrose land; primroses everywhere, not small isolated flowers but huge strong clumps, smothered in fine long-stalked flowers of purest yellow. Away they run up the wood in chains and patches, like a gigantic pattern of finest lace, nestling around the bole of grey lichened oak, encircling the sprouting stubs of pink-budded hazel and wych-elm: often with their delicate downy pink stalks hidden in dry rustling oak leaves and contrasting with the shining green rosettes of pushing blue-bell roots, or sometimes almost covered with green tassel-flowered dog's mercury.

What exquisite beauty is here—and why? Why this lavish waste? Are they displayed to welcome the first notes of the newly-arrived chiff-chaff, or are they there to hide the desolation caused by the greed of the flock of heavy-winged wood-pigeons, which have

PRIMROSE LAND

stripped the intermediate patch of wood-anemones (leaving a myriad of headless stalks with only leaves), indifferent to their feast of poisoned beauty?

Primroses everywhere! Nature's squandered abundance, each root perfect, and each in a different setting. Thus it would seem as if the first warm rays of early April sun had caught the face of the wooded bank, and with their touch had come a crop of golden freckles.

A BANK OF PRIMROSES

EARLY SPRING

I⊤ may be later March or early April days before the trees have unfurled their leaves ; cold east winds have forgotten to blow, and there is brilliant sun. We have seen the brimstone butterfly waked from its long winter's sleep, an orange tip scarce hours old floats by, and humble-bees in buzzing life are engaged in their wondrous pursuit of 'sensing' a cavity in which to form a nest.

The blundering oil-beetle (*Meloe proscarabœus*) with its marvellous legs scrambles over the undergrowth in the hedgerow. How strange that its errand should result in the weird entanglement which connects its life with that of the humble-bee! Naturalists tell us how the eggs of the oil-beetle deposited in early flowers hatch out to small, active six-legged creatures which, crawling on to the back of the visiting humble-bee, are carried by it to its home, there to lead a pilfering life by feeding on the bee's own eggs and, later, becoming fat maggots eating up the supplies of honey.

ORANGE TIP
BUTTERFLY
(photographed alive)

THE OIL-
BEETLE

66

EARLY SPRING

High up, a flock of fieldfares wing their northward flight, strange choice to flee from genial spring! No more their cheery note 'tchack-tchack' will sound till winter's cold returns.

The dawn has awakened a thousand throats to welcome in the day, but as the sun climbs upward bird labour begins, and noon finds them almost silent, for there is much to do; the work of selecting a suitable nesting spot and the gathering of the necessary building materials is so absorbing, that they have only time to utter broken notes and happy courtship calls; but, when evening comes, they burst forth into song again and close these joyous days of life with melody.

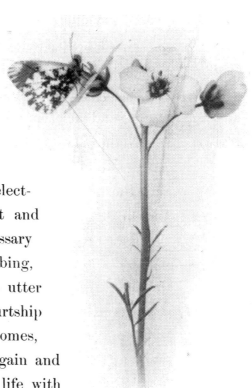

Unlike other birds, the lark seems unable to suppress his joy, and springing

ORANGE TIP BUTTERFLY
ON LADY'S SMOCK
(photographed alive)

up into the blue, mounts ever higher and higher and becomes a tiny speck which pours out a rapid torrent of song without a pause. I wonder why these shy retiring birds display such bursts of energy, for the combined effort of ascent and song must be very great, and what stage in evolution is served— scorning the tree or sheltering clod from which to serenade his mate, the lark soars aloft until he is lost to her and us, except for song.

THE LARK

Nothing now will check the onrush of spring with its wealth of flowers and opening buds; the primrose and the daffodil have dressed the copse-wood, and the marsh is clothed with kingcups bold; by every bank the lesser celandine displays its burnished yellow stars, the newness of its glossy leaves all glistening green. Even the tardy ash has waked, and the black winter resting-buds bursting, reveal within their soft tan brown scales rounded masses of purple-maroon anthers.

THE ONRUSH OF SPRING

Yes, spring has come and we can watch the birth of life, but know not whence it comes, nor why. And when we have greatly longed for it and counted on its joys, it is only human, I suppose, that there should

68

AN EARLY SPRING MORNING IN THE OAK GROVE

be a strain of sadness somewhere in its presence. Already the snowdrop and the aconite have lost their welcome charms, the crocus fades from sight and catkins wither brown. It is passing strange that when the waking call has sounded through the land, some should hasten off to rest and sleep the summer hours away.

Too well we know the hurrying race, too soon the fading petal falls 'and the grace of the fashion of it perisheth.' If only we could scotch the wheel, could stay the Hand of Time (that quickens as we age), could wrest its beauty ere it slips and leaves us but a memory!

70

FLOWERS OF
LARCH TREE

KINGCUPS

Down in the marshy meadow-land royal kingcups are ablaze, they gild the sluggish swampy stream, gold-lace its half-dried mud, and make the meadow stretch of lingering daffodils look pale, and the cowslips almost green beside their handsome clumps. The contrast of the deep rich green of their lovely leaves adds much to their beauty.

It is wonderful how Nature moulds the leaves of plants to suit the flowers. We should not blend the spiky blades of daffodils in all their bluish silver sheen with bold marsh-marigolds, nor place their rounded lobes beside the lilies of the field, yet each makes each complete, just where they are and as they are.

KINGCUPS

ROYAL KINGCUPS ARE ABLAZE

AN APRIL SHOWER

THINK of the drama of an April shower! A combination of comedy and tragedy. How suddenly the brilliant sun is lost, hidden by a bank of clouds which has quickly formed up in frowning blackness from an invisible source; gusts of cold wind follow in almost explosive abruptness; genial spring, which was with us a few moments ago, has gone, the counter-attack of winter triumphs, great drops of wind-borne rain are driven first one way then another and then fall in torrents.

Precocious young green leaves and shoots are lashed, bruised and beaten as the sudden storm roars by; the note of the rattling patter of rain-drops changes, the air whitens and a million balls of ice are hopping, bounding, jumping, as if a legion of fairies were holding a tennis tournament: from a thousand tiny hail-bows our eyes rise to the great prismatic arch above: on yonder hill the sun is shining. There is a lull, in a few gusts the trees shake themselves, like great monsters after a bath, and brilliant sun

73

warms the steaming earth smelling of growth : the
tragedy is over. Nature smiles on hosts of opening
buds, loosened by the rain and freed by the wind
of protective—but now needless—winter covering ;
under the beech trees we can sweep up the downy
pointed pale-brown scales, slow growth has made a
jump, spring is victorious, a chorus of happy home-
hunting birds shouts with joy and the tide of life
rolls on.

But in these days we are so busy that we only
grumble at the hindering weather, which, in its sequence of changes, is just perfect for the new birth and
childhood of a countless host of tiny seeds hidden in
the brown soil. Its magic key sets free the imprisoned
life locked in these minute specks, and the earth again
is green.

Thus sounds the waking summons to Nature's
'weeds' which hear the call, and the annual and
many another flower hasten to clothe and paint our
fields and gardens with endless forms of wondrous
beauty.

I know there are gardeners, I mean great and
undoubted plant lovers, who scorn an annual. How

THE
ANNUAL

much they miss! it is a great mistake to confuse rarity and beauty. They will rave over the blue of *Gentiana verna* (and rightly so), but is there no beauty in a blue pimpernel or the ubiquitous lobelia? After all, if I cannot grow *Eritrichium nanum* I can still admire and love the blue forget-me-not which grows of itself. Our views are sadly distorted when we admire only on account of rarity or an assumed (but perchance accidental) triumph of culture, and fail to fall down and worship before a field of golden cowslips or a bank of nodding thistles.

A BANK OF NODDING THISTLES

FRITILLARY

APRIL is the month of bulbs; chaste narcissi of unlimited variety, with wonderful recurring faithfulness, spring from forgotten nooks, clean and fresh looking. There are hyacinths and early tulips, scillas and muscari and many others that we love and the garden is ablaze, but few can compare in delicate grace and elegance with the common and easily grown fritillary. Thriving in damp or even water-logged positions, in grass or semi-shade, the graceful arching stems rise with rapid growth, carrying their dainty drooping angular bells. The mottled plum colours of the outside of the flowers are delightful, but it is only when we lift up the modest cup that the wondrous interior with its exquisite chess-board

SNAKE'S-HEAD

76

FRITILLARY
SHOWING TESSELLATED
INTERIOR

pattern is seen in all its beauty and colouring. Delicate and refined as the white variety is, it lacks the fascination of the markings of the more richly coloured types.

I often think that long, long ago when gazing on its checkered beauty an inspiration came to man,

For Art's unerring rule is only drawn
From Nature's sacred source.

From Nature's store, artists, builders and architects alike derive their first ideas, and, in the marking of this flower, or some of its many Eastern cousins, was

FRITILLARIA MELEAGRIS (Snake's-head)

found, perhaps, the germ of those marvellous mosaics and tessellated creations of old. Who can tell?

One cannot see the crimson glories of a sinking sun, when its gorgeous colouring is viewed filtering through the great perpendicular pillared boles and interlacing arches of tracery formed by the boughs of a high beech grove, and not be reminded of massive cathedral pillars and great arches, leading up to the groined roof, backed by the magnificence of its stained glass windows.

A GROUP OF IMPERIAL COUSINS (CROWN IMPERIALS, RED AND YELLOW)
IN THE BORDER

ORCHARD BLOSSOM

DOWN the grass-grown lane there is a stile and over its crooked rail an orchard where May has revealed a fairyland, or as if a tiny scrap of heaven had dropped from out of the azure blue and in the sunlight rested there awhile.

The apple trees are old and bent, all overgrown with moss and lichen, but half snow-clad in pink-tipped white, 'midst carmine buds; while all below the burnished buttercups gold-spot the rich green grass where daisies hide, and from out the hawthorn hedge the blackbird pours his liquid melody.

I wonder if there is in heaven's land a fairer sight, for, if there be, then must our hearts and eyes grow much deeper and larger: they are full to overflowing now; yet so must it be, for 'eye hath not seen, nor ear heard, neither have entered into the heart of man, the things which...' are there, and fade not.

THE BEECH WOOD

IT is a perfect spring morning in 'my garden' and the possibility of attempting to describe the scene before my eyes seems hopeless. Ah! could I but hold the artist's brush and steal one look from Nature's face, and with genius interpret her mood, when thus she smiles and speaks, even then it would be but a soul-less copy—dead—an incomplete expression.

For what words can portray it? words—the least impressive of things, hard and cold in black and white, with no life, whose familiar shapes and harsh sounds may suffice for the market, for the latest price of stocks and shares or squabbles of politicians; good, indeed, for concealing the true and real meanings of the heart, blinding the eyes and hiding the depths of life, but, for 'my garden,' this day of spring, they are futile, limited, lifeless.

QUEEN BEECH

For the call of the forest compels me and I see its Queen arrayed in matchless beauty. If the giant rugged oak is the King, the beech must be crowned Queen; she reigns here supreme, showing her

80

mightiest efforts and lifting a proud head a hundred
feet high on each of ten thousand clean smooth
rounded columns, arched and interlaced in graceful
curve and sweep, that lead the eye up and up to where
the netted filigree of tiny twigs would seem to brush
the dome of heaven; her charms are in her sweet

YOUNG
UNPLEATING
BEECH LEAVES

moodiness, her always fair uncertainty. Her lower
branches are almost in full leaf, clothed with those
fragile, unpleating, exquisitely beautiful pale-green
leaves, just sprays of giant maiden-hair; while her
topmost branches are still bare, and, though kissed
and tossed in gentle breeze and sunshine, have not

yet waked, except, as it were, to unveil their eyes by casting from their buds the silky coats, too small to clothe the rapidly swelling growth, as they unfurl their new-born crumpled leaves.

And so the life of each leaf has begun, its short existence of sun-absorption contributing its iota of help in rearing the perfect and superb elegance of the mature tree; and, passing through its sequence of greens to autumn's golden brown, its work all done, it looses its hold and floats down on its first and last journey, to add to the brown carpet, monopolising the ground and forbidding all save a few favoured flowers to spring up in the almost tidy setting of the pillared grove.

SHADE UNDER
BEECH TREES

Few things flourish in the shade and decayed mould of beech leaves, but they provide a home for patches and mounds of greenest moss, which, not content to weave carpet patterns, in some instances climbs the great boles of the monsters, following the tracery of branches to a great height, there giving place to greyer lichens where the sunlight glints and the myriad boughs dance in cross shadows of sun and shade. Through yonder glade the stems are

dark, the moss deep olive-green, while, just beyond,
sun-shafts break through and light up gnarled limbs
of silver-grey hung with huge sprays of
yellow-green leaves, so young and
delicately frail that they must hurry
along to thicken and grow tough
enough to withstand the rush and
angry blast of the mighty wind, which now in gentlest
mood lifts and lightly rocks them in the great avenue
of checkered shade.

YOUNG
BEECH
LEAVES

Yet all this in no way describes the extravagance
of beauty, for here the Queen of the forest holds her
court, and here she and Nature, unmolested by man,
are spending their substance with, it would seem, but
one aim—primeval beauty. For, look where you
will, the glades vie with one another, all are different,
yet nowhere is there any single glimpse that can fail
to charm, and with the soft murmur of the breeze in
the topmost branches comes the, as yet, not too
frequent note of the cuckoo.

Down that avenue are clumps of holly, their dark
polished green dress contrasting with the background
of rich brown beech leaves on the ground, and just

THE YOUNG UNPLEATING LEAVES OF BEECH TREES

beyond a tangled mass of once tall bracken, now crushed by winter's heavy snow to form a dry warm house, from which peers out a bright-eyed rabbit. Across the path there lies a victim of the storm ; its sundered prostrate limbs proclaim the mighty rending crash with which it fell and drowned the blasting roar of angry Nature, tearing and throwing up a wall of roots and earth, itself to lie unmoved and rot, the giant battered limbs, moss and fungus clothed, some so thickly coated that no sign of bark is seen ; a padded seat on which to rest and watch a shower of green leaves, rising and falling in drifts among the boles and branches, stippled and dappled by a thousand shafts of varied light and shade.

A VICTIM OF
THE STORM

DAYS IN MY GARDEN

The time of half-leaf to full-leaf, those days of May in woods and fields, when heaven dips to lift the soul of man. Oh! ye who jostle in the race to outrun time, to gain the prize which perisheth, unlock the shackles of your mind and go, walk alone, with May! Tread the soft meadow grass, where thoughtless feet may crush a thousand gems, behold the

> Summer-snow of apple-blossom
> Running up from glade to glade

and rest in the shade where bluebells haunt the leafing woods. Look where the beech boles lift their heaven-ward arms of zinc-hued grey, drooping like broken spray with their million hosts in new-born green ; forget for one brief spell the breathless blinding race of life, and enter the great galleries where God paints. Let the filtering shafts of sunlight touch the dust of life, and gild the dark places of your soul like the sun-patched woods of spring.

A MAY EVENING

It is a May evening, when the shadows lengthen and solitary rooks slowly and laboriously flap on their straight line of flight to the distant home wood, and

CATTLE GRAZING IN THE EVENING

the chorus of bird voices individualises, so that the continuous song of each is recognisable and its distinctive charm is heard; then it is that Nature is slowly wrapped in restful peace, and seems to make a pause.

As the twilight advances the wind drops and in the calm it almost seems as if the mysterious power of growth was stayed and the production of delicate leaf and flower and stem had ceased. They had material-ised so rapidly since sunrise, that we almost felt we had caught something of the mystic change, and seen the thin air and invisible moisture cohere to finest tissue, but now there is a halt —a quiet stillness—as if the current had been switched off, and all is silent restfulness and peace. Labour in the fields is over, for hay-time and harvest are hardly yet a promise ; there is no sound of cattle nor bleat of sheep, for they are again eagerly grazing the tender sweet grass, brushing on one side with moist cold noses the lovely scented cowslips with their pale-green heads and drops of gold. For these animals not having yet fully shed their warm winter coats have been resting in the shade and, like us, have felt unduly the first touches of a heightening sun.

COWSLIPS

88

HUNCH-BACKED BUNNIES
ADVANCE

A MAY EVENING

Bunnies from out the copse advance, the younger brown puff-balls with the recklessness of unwary youth, followed in cautious hops by their hunch-backed seniors, to whom however the joy of life is so great that they cannot occasionally suppress an aimless pirouette and capering scuttle; while from the oak trees above, our delight is enhanced by the smooth soft cooing of the now almost tame wood-pigeon. Thus Nature, full of bursting life, seems to pause, to hesitate in silent restfulness, in satisfying peacefulness, free from all stagnation—yet not a pause, for life never pauses, alas! it passes soon, like a flower, and is gone.

Yet this hush of twilight brings a quietness we hear; it is far from real silence; all around are many sounds, but they are tender sounds, which seem to calm and lull the mind to rest, to soothe the senses, as when we feel our grasp of life relax and welcome sleep replaces the toil of consciousness.

THE QUIETNESS OF EVENING

Unbroken silence is a death-like thing, a real thing to feel—too terrible! a desert world of sun and sand, soundless, lifeless, a tract whereon the wind is speech-less, finding no rock to answer back in moan or sigh,

THE DESERT

89

and, with its sand-laden breath, fashions the liquid
slope of dune or melts its curve and ripple—the deep
despairing silence of the desert—a place, methinks,
whereon the Hand of God has passed, but never dwelt
to bless.

AN ARAB'S PRAYER IN THE DESERT

CHANGEABLE WEATHER

THERE are some people whose troubles in life are so insufficient that they make up their burden with the never-ending trouble of the weather. It is always with us and provides an inexhaustible topic, it refuses to conform to the calendar and is a safety-valve for all the eloquence and wisdom which our nerve-strained, ill-treated bodies generate, something at which we can always grumble and which will not answer back.

For there are people who are actually unkind enough to say that the English climate is changeable, terribly changeable. When a native of Texas heard this charge his eyes dilated with envy. 'You would not,' said he, 'complain of a changeable climate if you lived in one where skies change not, and where from sunrise to sunset for week after week, sometimes for months together, there is no change, nor hope of change, nothing but a blazing, withering, frizzling sun. What would I give to live in a climate that was changeable; where there was even a chance of change!'

No gardener is worthy of that high designation

unless he is somewhat of a philosopher: he has learnt
long ago the truth in those lines—

> I'll take the showers as they fall,
> I will not vex my bosom:
> Enough if at the end of all
> A little garden blossom.

He has learnt more ; not being satisfied with the last
line he uses his energy in increasing or forestalling
the showers, and the many other little changes which
he knows will come, and by his skill and patience
transforms the 'little garden blossom' into a wealth
and abundance that feeds his eye and fills his soul
with delight and thankfulness.   There is more than
one trap to catch a sunbeam and more than one way
to outwit Jack Frost.

In truth this changeableness may spoil our new
hat, or make us wish that we had chosen watertight
instead of patent-leather boots, but these are small
matters, and who is to blame if we or some people
will run risks by donning dainty creations, be they
those of hatter, costumier or bootmaker, when pru-
dence and the barometer unite in proclaiming such a
course unwise?

For we soon tire of settled weather, three or even

two short weeks of hot summer sun and we are clamouring for rain, we are weary of the rattle of water-cans and cowl, and the coils of the hose have become snake-like and venomous; patience has gone. A 'spell of frost' or a foot of snow with a frozen water-pipe have almost enlisted us among the army of grumblers.

But let us think, it is to this so-called changeableness that a great debt of thanks is due, to it our country-side owes much of its beauty, it is the very life and source of its varying delight, and of ever fleeting tones of shade.

It gives us the rift of eau-de-nil in the banks of storm-driven clouds in shortening days of blustering wind and downpour, when leaves are falling fast, and the blazing glistening rays of evening sunshine burst through and reveal a dripping world transformed and illuminated by colour : the sodden limbs of elms, chrome-yellow-leaved, are traced in deepest black.

It flashes brilliance where pointed larches drop fine threads of gold, fires ruddy maples, gilds the silver-columned birch and burnishes the copper-browns of beech.

EVENING SUN AFTER
A WET AUTUMN DAY

93

It gives us the threatening stillness before the coming storm, the lights of leaden skies, the majestic

WHAT CHANGEABLE WEATHER GIVES US

magnificence of booming tempest and mighty roaring winds, the splendour of the fleecy floating cumulus cloud, the crisp pattering noise of rain on many leaves. Those driving mists, whose filmy wisps curtain the mountain steeps and mould anew their towering crests, and swathe their crags in a vast wind-borne veil of gossamer; then kissed by the downward sunlight beam—a golden bar dropped down through storm-rift clouds—melt away and fade to thinnest air.

EFFECT OF SUNLIGHT ON THE SEA

A thousand diamonds flash on the blue-green waves as they lift and toss uneasily, now lashed in heaving dulness and dashing their crested whiteness against the black wet rocks and hissing in anger at thus being kept in bounds. We see the slanting sun-shaft, like a mighty sword, thrust its gleaming brightness through the distant dove-grey sea fog, piercing its shadowy patches and, striking the sullen bosom of the ocean, reveal a shield, lighted with the fires of opal. Over the fertile valley, studded with a myriad of rounded tree-tops, cloud-shadows race,

94

rippling green to blue and grey; a shifting, moving scene, changing even while we watch, always beautiful, never the same.

'I wonder,' said the townsman to the lover of the country, 'that you do not tire of this view, you always have it.' 'Tire,' said the other, 'I never tire of it, because it never looks the same twice.'

The Australian bush in many parts has very much the same appearance all the year round, the seasons have little effect. It is monotonous and lacks those changes which create for us a daily, nay hourly, delight of colour and effect, in sky and land.

When the bleak dark sunless days of winter are with us, it is difficult to imagine the sunlit flower-land of leafy June, the change in Nature's face is so rapid and so stupendous, yet we should soon fail to appreciate an endless June. In all those days of longest light, from the saffron-flushed dawn to the emerald and sapphire night that knows no darkness, there is no real sameness, no two days are alike in sun and shade; we look upon Nature's ever changing beauty. The very storms that would seem to scar and rend her face are but the flux with which she

95

smelts her splendour, and her rapid changes are but her method of renewing her moulds in which to recast her perfect productions, reminding us once again that as the thread of life runs out and meets the tangling storm-winds of sorrow, so the unravelling skein re-weaves again those new beauties which are forged and wrought in the furnace of pain.

Nothing lasting, nothing strong, nothing beautiful, is known in God's Creation that has not come and is not conserved through struggle. We dare not overlook this truth in the management of men's affairs. G. A. B. DEWAR.

BLACK

NATURE exhibits her superb mastery in art when she takes from her colour-box the tone we call 'black,' but when she thus paints with lightless night she uses it very sparingly and endows it with a property that knows no dead dulness, in its freedom from all gloom, very different from the lifeless and uninteresting black attire which seldom becomes the wearer and is inevitably associated with sorrow. 'Why should I wear black for the guests of God?' asks Ruskin.

BLACK

There is a glossy brilliance in the coat of the wild black rabbit or healthy domestic cat, a polish and metallic lustre deck the fiddle beetle, and a score of other so-called black and peacock-blues and greens are sheened on the magpie's tail and wings, while the sombre rook is iridescent in his coat of changing colour. Often hidden, yet still there, Nature blends with the sooty markings on butterflies' wings rich purple blues, deep orange bronze with violet lights, or boldly breaks them up with dazzling scarlet and white, as in the wings of the red admiral.

SINGLE FLOWER OF BEAN

FIELD BEAN

There are no dead blacks. When her brush on flower-land dwells, with matchless skill she introduces the spot or line with marvellous attractiveness and, when more generous, with amazing beauty. What texture can surpass the velvet bloom-washed folds of the black pansy? while the black arum attracts by its weird uncanniness.

It is said that the blackest spot among flowers is the blotch on the wings of the common field bean, a strange mark and certainly dense black and very beautiful.

Whenever I gaze in dumb wonder at the superb collections of orchids often seen at our large flower shows, exhibiting the most exquisite colours in brilliant and delicate shades, in refined combination with the highest development—the aristocracy of flower-land—my eye never rests till it has sought out one variety which fascinates me. *Cœlogyne pandurata* is a perfect gem with a wondrous blending of contrasts. The petals, which have a fine crystalline dust appearance, are pale green, the green of unfolding beech leaves, while the exquisitely frilled green lip is spotted, laced and netted with intense black; it

FLOWER OF
CŒLOGYNE
PANDURATA

98

BLACK

is marvellous—but descriptions of orchids are futile, for they possess personality, and demand our love, and so should live for ever.

It was in September that I found beneath the trees a wondrous moth called Merveille-du-jour, whose lovely upper wings were like a scrap of Paisley shawl woven in apple-greens and silver-grey, and between the patches ran an angled thread of black with little V-shaped spots—but Oh! the wisdom shown when it flew and alighted on the bole

MERVEILLE-DU-JOUR
MOTH

THE SAME MOTH ON LICHENED OAK BARK

of a lichened oak, its beauty invisible in its perfect mimicry of the coloured ground-work of grey and green moss-covered bark whereon it rested.

Who taught this floating gem of life to steal the pattern of its mottled coat, and to seek a protecting spot of safety, selecting its host among all other trees around? Instinct? . . . Truly, some words should be writ in large letters, for they have large gaps to fill!

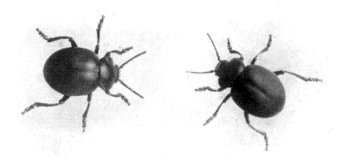

BEETLES (TIMARCHA TENEBRICOSA)
LIFE SIZE

100

BLUEBELLS

To-day the woods are bluebell-hazed and leaves are tenderest green, the awakened oaks are half-leaf clad, all catkin-hung, their tiny rouge tips are shading to yellow-ochre-green, almost transparent in their delicacy; a few trees are nearly in full leaf, while some have only just released their buds.

Somehow these pale new fresh greens are much more beautiful with the bluebell-blue than the darker richer greens of summer.

Masses of flowers when seen—be they a wood of bluebells, a field of buttercups, a moor of heather or a cornfield set with dazzling scarlet poppies—produce a surging wave of pleasure; but there is a something inexpressibly charming, bringing with it a feeling of the satisfying fulness of beauty, in the sight of the new-born greens of trees and woods in spring-time— from the time of half-leaf to full-leaf.

OAKS ARE
HALF-LEAF
CLAD, ALL
CATKIN-HUNG

101

As we grow older we have increased admiration for youth; the maiden's face may have small claim to beauty, but there is about it and her the indescribable charm of the atmosphere of youth, the untarnished bloom of the spring of life that comes but once; and I think there is that same atmosphere of charm when the maiden bloom of spring-land paints the wooded country-side in freshly created tints of green.

'Tis but an incomplete fragment we can place on paper, artist's canvas or the truthful photographic plate, of the beauty of the sunlit glade when spring-time oaks and fronded beech or fairy-feathered birch, half shade the sun-patched bluebell haunt where cuckoos call and answer, when Nature hides her cruel strife, resting in peace, and bids us see a heavenly land and worship Him who made it all.

> In Thy great out-of-doors!
> To Thee I turn, to Thee I make my prayer,
> God of the open air.

AN OAK TREE

A GIANT OF NATURE'S GARDEN

WHAT a delight an English oak tree is! What a
pleasure all the seasons through; what a possession
should it grow within our own ringed fence set on a

103

carpet of green velvet turf, and the number of its birthdays be a speculation !

Its great corrugated bole is a sight of which we never tire, with its rare combination of beauty, utility and strength ; and if we are in the mood to meditate and are wise enough to allow our mood to prevail, our oak will provide a mine of subjects, of unanswerable imaginings and speculations. As the soft June breeze gently lifts and sways the great branches we watch the responsive shadows dance and as it were 'set to partners' on the lawn, and as we gaze up into the mighty network of rounded limbs and rustling leaves, there comes a feeling of the smallness and limitation of our knowledge.

AN ACORN

For once, long since, was not the plan and habit of life of this complex mighty tree hidden within the embryo of a little polished acorn, that perchance narrowly escaped forming part of some wary old cock pheasant's breakfast half-a-dozen centuries ago ?

And yet how great the measure of our knowledge compared with that of the nuthatch, who day after

104

day runs up and down its branches and shyly peeps round their hidden sides; happy in his ignorance, he daily re-examines the same haunts, the very crevices of its crinkled bark (his larder and his anvil) are known to him. Where is his ambition? has he none? Why does he not fly to other woods and other trees and so explore the world? The world! the world of which he is a part and yet knows nothing. How absurd it seems that he should under-stand and realise but little of it, as we know it in its complexity, its pleasures, its glitter and gold and glamour, its striving after wealth, power and fame, its bickering and bargaining, its petty trials, its hells of selfishness and war. Of these he knows nothing, nor can ever even conceive of their existence in the sphere in which he plays his part.

THE NUTHATCH

GERMS OF GIANTS
WITH A THOUSAND
YEARS OF LIFE

Ignorant of the very tree in which he spends so much of his life: we realise his limitations and pity his ignorance; how much in life he misses! For do not we (the lords of creation) know all this, do we not know and understand the conditions which brought about the germination of the acorn and the laws by which its development has taken place? Can we not follow them step by step, give names to each process, long names too? We can identify the elements which have combined to build up the texture of wood, fibre, bark and leaf; we are familiar with their properties and the structure of cell and tissue, we can trace the results of a hundred different complicated actions. How great is our knowledge! and yet, and yet, how small, how little our understanding—

For we are but of yesterday and know nothing.

Have we not in reality to stand aside and just watch the working of never-failing laws, impelled by a cause we do not understand? ever progressing onward with an inflexibility of purpose to an unseen end, framed and planned by a Mind vast, infinite. Laws perfect in action and detail, silent in working.

Here is no throb of pump to force the life-giving sap along the maze of intricate channels, no erecting scaffold, no snorting engine nor sweeping crane, no noisy

hammer nor shaping chisel.   Where is the draughts-
man who plans the extension of arch and limb, or

'BREAKING' AN OAK TREE A HUNDRED AND FIFTY YEARS OLD

calculates the stress and strain of weights, thrusts
and foundations?   Where is the power house?   No
capital here, no labour of human 'hands' nor human

brains; no casting of moulds for leaf, limb or acorn cup, nor designs of bough junctions—yet somewhere the plan was drawn; the thing conceived!

And all through the complex structure runs, in every detail, the same wondrous law of beauty; even when the great fawn-coloured bole lies prone upon the saw-pit and is slowly 'broken' by the sharp hissing cuts of the saw, the solid planks reveal the marking of the superb grain, flecked with the flower of inimitable 'splay.'

A giant of Nature's garden! created and wrought by an Architect, the working of Whose Mind is infinitely further beyond that of the limited mind of man with all his boasted learning, than our small knowledge is beyond that of our little friend the nuthatch, who has just looked at me over the top of a bough.

Because for awhile we follow the Hand that guides and can bring to pass, because we name and partially comprehend the varied processes, because we are familiar with results, we are apt to persuade ourselves we understand the cause, or perhaps never even stop to think that a cause is necessary; and so our ignorance of our ignorance is not less than that of the nuthatch.

'TEMPEST'

UNSEEN, unheard, yet in the close calm afternoon we feel there is 'tempest' coming. It is not a question of the barometer, or the wind, certainly not the meteorological forecast—as yet there are no threatening clouds—we know it is coming, simply because we just feel it.

LATE MAY

The fragrance and beauty of the lilacs have gone, save perhaps of those valuable semi-double forms which bloom later, but the laburnums are still golden glories, and *Azalea mollis* in all its lovely and deli-

cate shades of orange-flame and sunset tints, fills the
still air with its perfume, while the tree pæonies
seem too lazy in the sultry heat to open their great
flowers; here is an immense pure white one whose
crumpled petals limply flop over and hide the large
crimson knob of its pistil, encircled by a gold dust-
laden fringe of anthers. One is reminded of the
Eastern maiden the flash of whose dusky eyes has
kindled our imagination as to what features may be
hidden beneath her white veil.

Yonder are splashes of lovely ruby-crimson, maroon
and blackish bronze, uncommon shades—they are the
last of the Darwin tulips—and as we wander on we
are suddenly aware of the stillness; the birds are
silent, no breath of wind to sigh or rustle and, lifting
our eyes from earth's floor to heaven's dome, we see
the blue-grey leaden-hued shapeless banks of clouds,
gathering in haste, massing for the approaching
storm; we dread it for the flowers' sake; for itself,
its superb and mighty grandeur, our words are idle.
On it comes, the moments now punctuated with distant murmurs that grow to rumblings. But still we
linger among the flowers, for now is the moment of

110

moments. I know nothing in all the varied lights of Nature to equal those vivid contrasts and transforming effects which often just precede the summer storm. We cannot see the sun shining, yet the light is strong and flowers and trees stand out in clear-cut brilliance. Whiter than snow, the 'Whitsun bosses' seem to hold their shower of falling snow-balls, the glaucous boughs of cedar and cupressus flash in brilliant silver-blue, metallic burnished lights, steely and grey, it is as if Nature now mixed her colours with the medium of polished metals, so lustrous are they, and the very elms are silver-coated against the frowning background of clouds; greens of every shade have become livid, the brilliance of all colours is intensified, while the distant hills have receded miles, yet they look higher. A flash! the rushing gust of wind, and now the booming bursts, and seems to crack the very roof of heaven, as answering echoes roll away. How can the immense heads of rhododendron, that wondrous 'Pink Pearl,' almost too heavy to hold up its own magnificence, or the oriental poppy petal, which seems to have such a frail hold on life, withstand the deluge? The all-too-short

LIGHTS BEFORE
THE
THUNDER STORM

111

bloom of a twelve-month is gone, yet, 'we must not vex our bosom' but be glad that the dry earth may receive a watering in a few minutes such as a legion of gardeners, armed with a thousand watering pots and stimulated by a thirst-quenching energy (which we always wish they possessed, but which they never do), could not accomplish in 'a blue moon,' as Old Peter says, however long that may be. And afterwards there is that smell, indescribable, but very good, and then—the time of toads!

A DAY IN JUNE

A DAY in June, a perfect day and yet I was sad—
unfettered Hell was loosed—men played with death
—sad news had filled the land, and so must it ever
be when we usurp the Power of God and thwart in
war His Will of Peace.

Out from the crest of the sun-kissed wooded hill I
looked, and all around lay stretched in greenest green
the wealth of leafy June, sun-bathed in perfect
peace, no strife to mar the rest, no jarring
note to mingle with the breeze that silver-
waved the meadow grass and set the leaves
a-dancing.

Around my feet was spread in deep
rich yellow hue the bird's-foot trefoil, its
brilliant buds just points of fire, and the
lazy small heath butterfly flopped carelessly
among the frail blooms of the pale chrome rock
rose.

I watched the grizzled skipper flit by on
hurried errand bent, the tireless swift with its

SMALL HEATH
BUTTERFLY
ON ROCK ROSE

circling flight, which seemingly has reached perfection, high up in the blue, and I gazed afar over fertile green until the patchwork of the hedgerowed fields and wooded slopes was blue mist-washed; and on

ROCK ROSE

and on from blue to blue to where the distant hills rose up beneath the straight flat base of puff-white clouds, and marvelled once again at Nature's skill in blending thus her blues and greens, to summer-dress the land—peace wrapped in promise.

114

A DAY IN JUNE

And so the vast, the perfect restful beauty of God's
Garden spoke to me and bade me look afar and see,
beyond the veil, the land
where strife is not, nor
death nor pain, and
sorrow is unknown.

GRIZZLED
SKIPPERS
(photographed alive)

MOONLIGHT

I⊤ is of course only natural that we of the country-side should appreciate the moon more than the townsman. Eyes that rely on the street lamp lose their cat-like power, and are blind to many a lone track, easily visible to those who are used to the country at night. The quiet silver radiance of a moonlight night often makes one rebel against the demands of sleep, but the energy spent in the day that is past, and its recuperation for the next that will all too soon hurry upon us, are conditions which must be met. But who is there in those all too infrequent times, when the cup of pure life holds health and happiness, that has not been saddened, nay, almost angered, at the thought of how much life is seemingly lost in unconscious sleep? Sleep—thou life robber! thou monster of interruption!—snatching me away from all my desires and joys, yet in thy kindness and by the miracle of thy power renewing the subtle threads of life, and by the magic of thy silent hand holding me secure, unknowing, on the very brink of infinity. And so, thus blessed, we miss the sunlit Queen of Night who with her silver sheen of beams

116

MOONLIGHT

has kissed yon mountain top, and with unbending rays blacked out its jagged ridge and made of it a silhouette. Her colour-box has limits, for where the midday sun has shown us rainbow tints and verdure green, *she* paints in black and white with sombre shades, and the distant ridge of softest blue and green in daylight, is sharp and hard with greys and black by night, and clear-cut lines mark where shadows fall into darkened deeps of mystery.

But land is not her canvas, save perhaps the mountain crests; she loves to paint the sky and sea and silver-wash the stream. Look where her brush in passing down has lightened up the silver-rim-tipped, cloud-flecked dome of heaven and sent its arch of deep steel-blue far back; away beyond the place which it in daylight bounds, and we seem to see *behind* the stars of glittering gold; those countless worlds; 'the pulsing dome,' unthinkable! See where they hang, and yet beyond, in depths of ether's blue infinity. How vast the circle of her path seems, how small are we who watch her noiseless ride in space, we, whose tiny atom of life is as a grain of sand, a sigh of wind! We feel a crushing sense of our little-

117

ness when we think of all the eyes that see her brightness now, of all those who have looked upon her face
in æons of the past, while she with undiminished
light, and unfailing regularity, still shines on.

> Unto His measures
> Moveth the whole

beyond the ken of man, beyond his scales and rule.
And we are glad to come to earth again, all dusky
grey and shade, save where the moonlight beams have
touched the sleeping lake and turned its darkened
surface into molten lead, catching the eye of water-
fowl and staying their weird night-flight to rest, or
further on, where the river's bend has whitened to
quicksilver; to gaze upon the long straight wake
looking like moon-fragments, as on the ocean's
heaving breast they lift and dip and dwindle in a
tapered wedge—a glittering path—to meet the sky.

MOONLIGHT
ON THE SEA

> Thou gavest me wide Nature for a Kingdom
> And power to feel it, to enjoy it.   Not
> Cold gaze of wonder gav'st Thou me alone,
> But even into her bosom's depth to look,
> As it might be the bosom of a friend.
> The grand array of living things Thou madest
> To pass before me; mad'st me know my brothers,
> In silent wood, in streamlet, and in hill.

118

JUNE

June is called the 'month of roses' but now, thanks to much patient labour, we have roses, exquisite roses, from May to November.

'Tis true you cannot paint the rose, nor gild the

A WOODLAND PATH

lily, yet some have enticed Nature to re-arrange her colours and repaint her blooms, to refine their dainty curves and mould anew creations of beauty, undreamed of before, and those who have thus gently coaxed her in her moods have drunk a cup of sweetness.

119

But June brings more than roses. Nature's ban-
quet is now spread in garden, wood and field, her
bounteous table loaded with flowers
strewn in lavish profusion.

If we had lived a winter-life and
never known a June, if our eyes had
never seen the summer's green, nor
looked on speedwell-blue, and then
one day had waked to tread a woodland path, and there beheld the
green-leafed world, the forest's
summer dress, had smelt the wood-
ruff-perfumed air and brushed aside
the bracken fronds all russet-antler
topped, to where the meadows

lie bejewelled in
the sun; methinks we
should have thought
we'd waked in heaven,
nor would the truth
have dawned upon us
as we walked through
the meadow grass and

TALL BRACKEN OF THE WOODS
ALL RUSSET ANTLER TOPPED

120

with wonder-wakened eyes looked on its radiant robe.

The crimson globes of clover nestle near the trefoil gold, the gilded tiers of rattle rise half up the buttercups, the snow-fringed ox-eye waves above the modest blushing daisy stars, the ruby-reds of sorrel seeming to rob the sinking sun, that glistens on the polished bronze of dancing quaking grass, the spotted orchid's lilac shades, the milkwort's deep rich blue—a rainbow-jewelled throng—the miracle of June!

I close my eyes, my mind has spanned the bridge, and I am back upon an Alpine upland pasture, knee-deep in *Anemone sulphurea.*

The blues of heaven stoop to meet the blues of earth, they mountain-mingle where the soft white clouds are pierced and torn, half-shadowed, almost grey, against the glistening whiteness of eternal snows, that look as if they were the very steps of heaven.

How can I forget the sight of those pale sulphur-coloured windflowers, gently bending in the breeze as if a great stately dance were taking place, having for partners their white sisters, in this spot less common than the yellow! I had no words to describe their beauty as the white curtsying heads revealed on their undersides touches of almost metallic blue: their grace, their charm was perfect. I stepped with care on upland bog to avoid crushing a host of *Primula farinosa*, butterworts, *Gentiana verna, bavarica* and *acaulis*. I climbed the col and in the mountain torrent-bed of pebbles found great round cushions of *Silene acaulis* and in the wet moraine *Ranunculus glacialis*. A legion of saxifrages, starred mossy things, some encrusted as if with a summer hoar-frost, and strange sempervivum, all cobweb spun, that I am sure could never please the neat housewife. A rock garden of mad dreams—a land of gardens gone wild! Nay, it is but 'Nature in her Divine purity,' utterly beyond description, impossible to compare with our English fields and hills. On such rambles I am always most conscious of one thing, an overwhelming feeling of weakness, the weakness of exhausted admiration, a sense of insufficiency of appreciation: beauty too great to grasp.

Again I span the space. At home I see the English

JUNE

flower-land, the fertile vale, June meadow hay.   Blue
smoke coils up from the half-timbered
thatched cottage, into the tall elm
trees; there in the haunts I know
well are the flowers of my childhood, nay more—of my own blood
they seem, for they have grown
into my heart and life: I love
them more, far more than all the
gems in yonder lands, because
they are my own, my very own,
and speak to me.

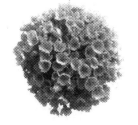

COWSLIP BALL

NOISE AND WIND

THE UNPLEASANT
NOISE OF TOWNS
ONE of the greatest charms of the country-side is
its abiding quietness, and a departure from it to the
terrible noise of the busy town is an agony. The rack-
ing roar of traffic, the harsh street cries and rasping
tones—when all that is unpleasant in the human voice
is brought into prominence—the confused and jarring
clamour of city life, form a terror unknown and never
realised by the habitual town dweller.

There are many sounds (noises some would call
them) in the country, connected with work on the

124

land, but they are mostly musical noises, their environment would seem to have softened and modulated them and they have become a part of it.

Hear the rhythmic swish of scythe, the clattering throb of mowing-machine, slaying its swath of innocent beauty, the droning hum of thrashing-machine ebbing and flowing, the cowman calling home his herd, the ploughman urging on his straining horses, the woodman's axe, the blacksmith's ringing anvil, the hundred sounds of fertile country-side, floating through the air ; all have exchanged their harsh and hurrying clash for Nature's sweeter tones.

THE SOFTNESS OF COUNTRY SOUNDS

MOWING-MACHINE

WIND

In the town there is nothing but unpleasantness in the dust-laden wind, it has no music, only an eerie howl or whistling moan; but out among the wooded hills it has a thousand murmured notes, its unseen hands attune a mighty orchestra.

Listen where its passing breath vibrates the needles of the pine in happy sighs, the gentle caress as it 'bloweth where it listeth' through the leaves and draws forth a host of rustling notes as in varying

125

humour it rocks and lifts the dancing boughs; its hurried, pauseless song, as through the stunted furze

BY WINDSWEPT THORNS

and wiry grass it rushes madly o'er the hill-crest and sweeps along the curving ridge by windswept thorns.

Down in the meadow grass it croons among the waving bents and frets them into curling ripples of

126

lustrous land-waves, swishes through the bending corn as if it were clad in silken robes. And when twilight stillness falls on the marshlands where stand the great breadths of reeds, its noiseless breath sways their drooping purple-brown plumes and rubs from the stiff, straight stems and harsh leaves a short, dry, crisp rustle; then, passing on, ruffles like an opening fan the glassy water at their feet.

Across the wooded ridge its quickened notes rise like an anthem swell, then melt in many a soothing moan, to fade away in soft and gentle sounds as breaking billows on a distant shore ; again to rise and sweep athwart the vale in a mighty rushing roar, and die away in a diminuendo. And in the silent stillness every leaf seems to stand and pause, and listen in the calm to catch the distant rumble of the summer storm, the first faint stir, the noise as of wings, and then the running patter of great rain-drops as they fall.

Inland, the wind is seldom cruel, but sometimes it seems to try its strength and, half in anger, lay the forest giant low that it has gently rocked and tossed since saplinghood. It is only out upon the open sea that madness seems to reign; unchecked its course, it scorns control and, wildly rushing, shrieks in cruel power, in savage play throws the mighty waves of darkness on the blackness of the rocks, to break like banks of blizzard snow in fractured whiteness, and, howling on, heeds not their hissing scream as they unite again, unbroken still, while we are left to wonder 'whence it cometh, and whither it goeth.'

THE LANE'S TURNING

THE path in 'my garden' leads down a winding
lane, uneven with cart-wheel ruts in the stiff red
soil—real red—that flavours the hop and lends its
colour to the apple. The air is full, almost heavy
with the scent of meadow-sweet; the lovely fluffy
cream-coloured heads, with their rich red-brown
stalks, are dotted everywhere amongst the young
growth of pollard alder stubs, mingled here and there
with patches of evil-smelling hedge woundwort, with
its peculiar stiff angular spikes of dusky purple-red.
Half round the turn (and the more turns these lanes
have the better) amongst the meadow-sweet, there
stands a steepled colony of noble foxgloves, great
tapering spikes of softly-shaded rosy blooms, a string
of finger-stalls, just made to fit the pudgy thumbs
of little hands, as yet too small and plump for work ;
and, of sizes too, to fit those tiny fingers : pink and
soft to match them. Erect they stand, a stiff-necked
generation indifferent to the gentle breeze which

FOXGLOVE

sways the plumes of meadow-sweet, and makes them dip and bow as if in mock obeisance to their stately presence. There are tall green arching grasses, breeze-bent, and great bushes of dog-roses, with a pink all their own, thickly set with thorns; a safe retreat for the shy black-cap who out-pours his rapid torrent of song. He alone seems in a hurry, for *we* would pause and grasp the placid scene, till we have drunk in its sweetness, stored up its many messages, ere the day when song shall cease and the rose petals fall.

DOG-ROSE

MEMORY ECHOES

I OFTEN marvel how it is that the mind is able to call up a whole sequence of associations, all the details of which are flashed before the inner sight, by the sudden appeal to one of our senses. The fragrance of some flower has carried the memory back a span of years, and a fragment of life is lived again: perhaps a day of happiness when we drank deeply of Nature's beauty, or perhaps a time of sorrow and despair, when cords were snapped and gaps were left, never again to be filled.

Involuntarily, the minutest details are clear to the mind's eye. How the sun shone that day when love was young and sweet and good, and I walked beside her; every line and poise a joy to me as she stepped from stone to stone across the rushing stream. How clearly I see and hear the hurrying rush of the water as it dashes over the great water-worn stones, falling in goffered frills, then throwing itself up in boiling whiteness, swirling against the rich red earth of the bank, glistening and sparkling when the sun-

THE STREAM
SIDE

light strikes its broken surface, or stealing off almost sulkily into a shaded backwater as if ashamed of its gloomy colour and loss of life. I see the green, wet, mossy boulders (patched with ochre and grey lichens, crowned with tufts of waving, bending grasses and the spray of frail blue harebell) resting, all unconscious of the day when a mighty rushing flood shall dislodge them and hurl them on in its sweeping torrent. I remember what she said, and the look in her eyes, as she paused by the leaning oak, whose network of flood-washed naked roots still grip the streamside bank in defiance of the undermining waters. Ah! how the sun shone then! and we were thankful, *then* for happiness, *now* for sweet memories and unbroken faith.

Scent is not alone in its power of thus recalling the days that are for ever gone, somewhere there is a subtle connection between fragrance and music. A few bars of some melody—we hear the old familiar notes and the vision is summoned, the secret door of a dim recess in the brain has been opened, and we see it all again; the current of thought is switched far, far away, and there, mirrored in accurate detail, is a once familiar and loved scene, for ever past, but for ever indelibly photographed on the plates of the memory, retained, pigeon-holed as it were, in that marvellous store-cupboard, suddenly rising and filling our mental vision, in response to an outside call, un-aided by any conscious effort of ours; the whole scene re-enacted, in which we take no actual part, yet in which we are the stage, actors, audience and theme. How little we know ourselves, how like a switch-board we are, subject to a thousand outside currents and influences, almost a plaything in the hands of operators by whom we are controlled!

So it is when the hot sun shines on those quiet and secluded old-world spots, where narrow paths are fringed by neat box hedges, and those miniature ribbons of green fill the air with a peculiar faint odour. In an instant I am back across the years and, conjured into vivid view, an old garden is seen. Voices, loved voices, now long silent, speak; faces, well-remembered, smile. The snow-white lace and cap upon soft and neat hair, the silken gown, the droning hum of bees, the unnamed rose of ruby red which clambered over porch and arch, the beckoning

BOX
HEDGES

133

pool, the huge square arbour of yew, with its un-
inviting bench and colony of spiders, the great straw-
berry beds, the dead-ripe fruit in aggravating
abundance, the cherry trees just across the lane,—a
score and more of boyhood memories arise—ah me!
what days! Methinks that then the world was kinder
—and strawberries were sweeter.

> To some a fragrant perfum'd breath,
> Just sweet—and that is all—
> To me—love's deepest words it saith,
> Its hallow'd scenes recall.

WOODS IN JUNE

In the hot days of June when the fields are a
shimmering haze, how delightful it is to seek the
thicker shade of the woods and be refreshed by their
cool dampness and absence of glare, but one cannot
fail to note the bounty of leafage which, in its wild
and utter profusion, is so extravagant that beauty
seems to have almost smothered itself.

The spring carpet of flowers has disappeared and
the scene of its lovely freshness, which filled the
atmosphere with varying delicate fragrance, has now

134

become the grim and silent battle-ground where the fierce struggle for existence of the mixed and tangled undergrowth is fought to the death. In the attempt to smother one another all kinds of devices are used, victory being with those which can obtain the most light and a sufficiency of moisture. How often we are reminded of the trueness of Mr Dewar's suggested motto for the whole green world— 'Strength Through Struggle.'

Sparsely flowered trails of woodbine of the palest yellow-white straggle over the brushwood, while here and there rise the chaste ivory spikes of the fragrant butterfly orchis, with weird,

BUTTERFLY ORCHIS
(HABENARIA CHLORANTHA)

WITH WEIRD, TINY, GHOULISH FACES

135

tiny. ghoulish faces hidden in their flowers ; yet some of the clearer openings are still gay with rich pink campion, and beside the damp path creeps the lowly yellow pimpernel.

Pale spotted orchids are there with their brown-blotched leaves, but these, like the honeysuckle, missing in the shade their share of sun-paint, lack the rich and varied lilac tints with deeper violet markings, seen on a host of their fellows, whose pyramids arise in the uncut meadow grass, outside the wood.

So also the ladders of quaint twayblade are greener and their flowers have longer lips here in their sunless bower than their more sturdy and brownish-tinged brothers of the open fields.

The fascination of many of our English orchids is increased by their uncertain habits ; vagrants of beauty, they appear and disappear for reasons known only to themselves ; we seek in vain where we have erstwhile sought and found them.

TWAYBLADE
(LISTERA OVATA)

136

WOODS IN JUNE

But three short months ago this selfsame wood was almost bare and cold, a fretted maze of boughs and twigs, all grey and brown, but no less beautiful. Growth—the coming-of-the-green—has transfigured its face, changed it in a way we calmly accept, though in reality the stupendous miracle of visible creation has been wrought before our eyes; for, truly, three months are but a tick of the clock of Nature.

But woods for winter days—to-day we crave the air, the freedom of open upland downs, the garden of the hills. Beyond where golden broom and the great cymes of the elder flowers sun themselves, 'twixt the new green bracken patchwork, there pause and rest; where the fine grass is dry and soft, the home of the little people of the highlands, short in stature, sweet in perfume, brilliant in colour.

UPLAND HILLS
AND THEIR FLOWERS

137

FROM WOODS TO UPLAND DOWNS, THE GARDEN OF THE HILLS

138

Here dwell tiny Maltese crosses of yellow tormentil, the fine white lace of the heath bedstraw, the frail crumpled gold of new-born rock-rose, soon to fade, daintiest harebells swaying in the breeze, the golden claw of the bird's-foot trefoil, and,

THE BRACKEN PATCHWORK OF THE HILLS
WHERE GROWS THE YELLOW TORMENTIL

139

everywhere we walk, the rose-red thyme with bruised sweetness, filling the air and taking our minds away to gorse and breezy heather hills sea-girt, and making us see the things which are not there, but which live in the memories awakened by its perfume.

'ONLY OUR CLOSE-BIT THYME THAT SMELLS LIKE DAWN IN PARADISE'

140

STRAIGHT LINES

IT is difficult to make straight lines beautiful
and it is but seldom that Nature uses them in any
considerable number or length, rather she loves to
bend and curve or effect completeness in the circle
and the sphere. The apparently straight lines of the
stems of the firs and larch are found to have a graceful taper, furnished with their concave branches of
sweeping curve. From the nodes of the scarcely
lessening bamboo stem break clouds of delicate light
foliage, obliterating all straightness, and there is an
enchanting beauty in the exquisitely plumed head
crowning the straight palm-tree bole, making it a
unique and perfect picture.

Think how Nature paints the summer sky of blue,
pale blue—for winter skies have deeper shades and
richer tones—that frames the spreading downs,
which rise and fall to dip and fold, to where lies
the land's crumbling fringe of bold grey rocks, lapped
by the waves of the sea, ever restless, yet ever
beautiful.

THE SEA SHORE

141

The sea's scintillating surface always gives a feeling of security in the quiet strength of the low hill ground, whose cliff-face holds its all-devouring power within bounds. It is this distant view from land of the moving ocean, eternally restless, which seldom fails to give a feeling of mental rest and calm—a paradox indeed—and it is here also that when Nature uses her one long, seemingly straight line, she does this with the minimum of simplicity and the maximum of effect, so great is her art. What can surpass in beauty the clearly defined line of the sea's horizon? What more simple than that thinnest edge where blue meets blue and nothing more?

NATURE'S ONE
STRAIGHT LINE

A 'SPOIL BANK'

By yonder stone quarry there is a spoil bank where the farmer's enemy the lilac-coloured creeping plume-thistle (*arvensis*) has taken bold possession, carrying out a colour scheme with masses of the lovely purple-blue flowered tufted vetch (*cracca*). High rose-purple-crowned spear plume-thistles (*lanceolatus*), that intoxicate the humble bee, are holding up a thousand cobwebby globular heads each arrayed with its silver-green armour of spines spaced with mathematical precision, and pointed so sharply and truly that the finest needle is a blunt and pitted spike in comparison.

There are patches too of yellow hawkweeds in variety, bending grasses amid stately cow-parsnip and great handsome docks all laden with their tiered heads of fruit; some bright green, others all rusty red and deep rich brown.

In the hot noon-day sun it is a humming, buzzing

HUMBLE BEE
ON SPEAR THISTLE

143

world of bees, while to and fro in rapid flight with
constant sharp turns fly small tortoiseshell butterflies, unable, seemingly, to
decide whether the musky scented
blooms are preferable to the heaps
of stones on which they bask and
wave their gorgeous wings.

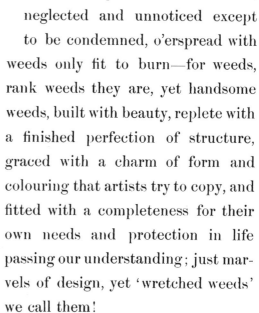

Just a piece of waste land,
neglected and unnoticed except
to be condemned, o'erspread with
weeds only fit to burn—for weeds,
rank weeds they are, yet handsome
weeds, built with beauty, replete with
a finished perfection of structure,
graced with a charm of form and
colouring that artists try to copy, and
fitted with a completeness for their
own needs and protection in life
passing our understanding; just marvels of design, yet 'wretched weeds'
we call them!

SMALL TORTOISESHELL
ON CREEPING THISTLE

Thus have we formed the habit of drawing
our arbitrary lines, of placing in water-tight

144

compartments all the things of life, dividing even those around us ; we wall up ourselves and them in prison cells, and bar out the beauty which is everywhere, but which we will not see.

CLOUDS

By the height of the cloud that sails, with rest in motion,
Over the plains and the vales to the measureless ocean.

THUS Henry Van Dyke speaks of clouds—'rest in motion'—and there cannot be a more simple or more perfect description. The grandeur of the mighty multi-domed cumulus clouds, their snow-white masses smoothly floating and gliding through space, seem in their calm and noiseless journey to be at perfect rest.

How much many of us miss by our unobservance of the ever-changing beauties of earth's dome. Cloud-land ! the abode of storms, the home where tempests slumber and awake, is furnished with a ceaseless variety of majestic ethereal forms, rayed with the brilliance of colours mixed with the medium of light, as they refract the smile or frown of the never-failing —if hidden—sunshine.

One summer day I watched the silent race across the azure span, watched where a thickening nothingness came and took a form, and grew, and whitened out the blue, then saw the glistening snow-white rim grow thin and grey, the wispy shadows of the filmy threads dissolve and vanish into blue again. Before my eyes they came from nothing into something and were gone to nothing once again. And more than fancy seemed to speak to me and tell me (what within my soul I knew) how some day, when life has freed itself from its cumbrous case of clay,

> in God's good care
> Somewhere within the blue,

we too shall come and go at will, our spirits ride at rest to far beyond the uttermost limit of that blue home, and there begin to solve in part the mysteries of the Infinite.

SIGNS OF AUTUMN

SIGNS of autumn are not wanting in summer days, if we care to look for them, but we are never anxious to note their appearance, any more than we are willing to realise the meaning of those tiny but unmistakable signs of failing, in the white heads and wrinkled hands of those dear to us. For then we see that which we do not wish to see, we feel a stab, a thrust—something within us shrinks; vainly we hope and try to think it is not so. Alas! in that lightning flash we have read the whole page, and though we hurriedly turn the leaf we cannot forget.

EARLY AUGUST

But these forewarnings of autumn should not sadden us, Nature's harvest is a time when she yields up her wealth, and, if she pauses after her bounty, it is but to rest, to sleep and rise the stronger to give again yet more.

And so 'twas summer still when we saw odd yellow crests that flecked the rounded top of many a hedge-row elm; while sheltered underneath were the brilliant polished red-berried columns of 'Lords-and-Ladies,' a food for birds but poisonous to man. The rowan tree was hung with its bunches of rich orange-red fruit, already tempting the thriftless black-bird and missel thrush; while fronds of gold and bronze-brown were showing among the ever beautiful bracken and flaming-scarlet touched the bramble leaves. As we stooped to gather the exquisitely delicate hare-bell, and smelt the almost sickly perfume of lady's bedstraw, we knew that the riot of summer flowers was over, and that Nature before casting her mantle of green was transforming her leafy threads and weaving gorgeous autumn glories.

HAREBELLS

148

OBSERVATION

By neglecting to cultivate and train our powers of observation in every possible way we lose much in life. Life however has made it necessary for us all to become somewhat skilled in the observation of those people with whom we come in contact. It is a marvellous thought that the combination of two eyes, a nose and a mouth in the setting of a human face can have such infinite variety, yet so highly developed are our powers of observation in this particular study that we instantly recognise and distinguish the minutest differences. And, what is still more wonderful, no sooner have we become acquainted with the individual face than its distinctive features become indelibly recorded on the plates of our memory, and though we may not be able to revisualise it at will, or, when we see it again, link up its connection with the past, yet our observation is so expert and the record so true that we are able to detect if any change has occurred.

So the specialist concentrates his whole powers

of observation on a selected object, and becomes an expert; in a flash he has seen and noted differences which we are only dimly able to recognise when they are pointed out to us. The Dutch bulb-grower Voorhelm is said to have been able to distinguish more than a thousand varieties of hyacinths merely by inspecting the dry bulbs.

To our eyes a flock of sheep is just a repetition of one pattern; not so to their shepherd: he will know a large portion of his flock by their faces alone, while to him a score of other characteristics make it easy to recognise each one individually.

If, as we ramble along the country-side, trees are just beautiful trees, and butterflies only pretty butter-flies, the grass and moss just a green covering, and the pond but a pool, how terrible our blindness! None of us can have failed to notice some difference in the birds or in the colour and size of the flowers, their variety is too marked to be overlooked.

It is only when our wakening observation is sharpened, and we seek to get a look at the face of the fly that settles on our hand, to understand the meaning of the tell-tale ripple on the surface of the quiet pool, to detect the half-hidden leaf of rarity, and note the warning calls of teeming life around, that our quickened powers open the gates of enchanted worlds, and as they stand ajar we catch a glimpse of vistas that have no end : it is only then that our eyes begin to see.

For then we realise that the world is a vastly richer storehouse than we ever dreamed or imagined, that every tiny crevice is filled, not with rubbish, but with the finest wrought work, that the mystery and complexity of matter is infinite.   Our newly-opened eyes reveal to us that on every hand there is limitless variety, the closer we look the more vast the number of differences.   No longer is the dog rose a simple dog rose, nor the bramble only a bramble, they are individuals belonging to great races, families and types ; we learn their faces, and where once all seemed alike, now no two are seen to exactly resemble each other.

NATURE A STOREHOUSE OF INFINITE COMPLEXITY

Hidden below the surface are Nature's moulds, which know no settled pattern ; casting with her restless atoms, she builds for ever anew her structure of beauty, progressing to an unthinkable end.

151

EARLY AUTUMN DAYS

EARLY SEPTEMBER

EXQUISITE and moving as the first days of spring are, when the re-awakening of life in Nature seems contagious, long and satisfying as are the days of summer, there will always remain a subtle beauty and richness about those fine hot September days,

when Nature yields of her mellowed fulness, days when she is finishing her task, filling her storehouse, ripening and maturing her growths for the sleep of coming winter.

There is a more marked contrast between the sun and shade, with a brisk fresh keenness in the air, yet without cold, unless perhaps an unusual energy has persuaded us to seek an early basket of mushrooms in the valley, where still lies the shallow filmy morning mist, and all the herbage is drenched in soaking dew, each step breaking as we walk a hundred gos-samer threads, all strung with fairy dew-drop lamps. Soon they will dis-appear, save perhaps on the shady side of that high hedge, where hang the great round webs tethered to

THE GREAT
ROUND WEBS
WITH THEIR
LOAD OF
DIAMONDS

their moorings by silken ropes, now strained to break-ing point, with their load of sparkling gems. For now the sun has flashed away the mists, in thrice gold beams of mellow light he floods the vale; methinks

153

more golden in these days than any others. It may be due to the absence of haze that, on these brilliant autumn days, the greens are yellow-greens, and the browns are golden-browns. Perhaps the sight (one of Nature's grandest masterpieces) of a field of waving grain, 'ripe unto harvest,' or of those unique and fascinating rows of sheaves ('stucks' we

'STUCKS'
OF CORN

call them in the west), spreads as it were a golden sheen o'er all, and gives a glory to the latter harvest days; all too short and few.

Sometimes it seems too sad that Nature in all her full-grown greenness and teeming insect life should now halt, that the same warm touch that waked her from her sleep and wooed her into such productiveness, should bid her cast aside her mantle of green and lead her into the silence of rest and sleep—a sleep from which some of her children arise and

154

again take up the thread of life, while others can only pass on the task to their successors.

Gay borders that had grown almost dull and dreary are now ablaze again, soft tones are almost gone and the pure yellows of perennial sunflowers and rudbeckias encircle keen dark eyes.

> I wander down the well-kept sward
> Which gallant Phloxes grace and guard,
> And while their autumn scent is sweet
> Gaze on the Rainbows at my feet;

their masses of rich lovely colours undimmed by great scarlet gladioli as they

> Unplait their green and break to flame,

where early aster blues gather together, sentinelled by stately clumps of the old 'red-hot poker.'

There is a wealth and fulness of colour in these fine autumn fellows, who have been busy all the spring and summer husbanding their strength, and who run riot in their splendour, displaying their gorgeous hues. Somehow they seem to realise that they must out-do the spring and summer blossoms; then, the violet seemed rich enough to satisfy—now, our appetite needs a mass of purple Jackmani; then, a tuft of

155

snowdrops—now, a colony of *Hyacinthus candicans.*
In these closing days we are colour-greedy, and are
only appeased by banks and masses, we need a whole
mountain of purple heather, a wood of flaming red
and gold. So Nature it would seem, all through the
long summer days, stores up in her verdant robe the
falling rays of sun-paint, until she can no longer retain
their hues, and bursting forth in her gorgeous 'fall,'
gives not only enough, but a colour-feast with
abundance to spare as she holds out to our view the
marvellous glories of her changing autumn gown,
woven with the very threads of sunset-gold, em-
broidered with its flaming fires and gemmed with a
myriad of diamond dew-drops.

> A garden is a lovesome thing, God wot!
> Rose plot,
> Fringed pool,
> Ferned grot—
> The veriest school
> Of peace; and yet the fool
> Contends that God is not—
> Not God! in gardens! when the eve is cool?
> Nay, but I have a sign;
> 'Tis very sure God walks in mine.
> T. E. Brown.

TOADSTOOLS

BENEATH the shade of the fine old row of elms where the ground is dusty dry, there is a wooden fence that has long since taken the place of the hedge, which in its early days formed the young trees' protection, but whose existence, for lack of moisture and light, they in course of time made impossible ; for so runs the never-ceasing stern struggle in the battle of plant life. The dark soil is enriched by countless layers of leaf mould and decayed boughs, and forms the home of many exquisite fungi.

AGARICUS RACHODES

Growth in all its forms is a mystery, and familiarity with its marvels deadens our appreciation of its

157

wonders. Hardened we must be if we are not amazed at its power and rapidity, when we watch for a few days a group of toadstools.

As if by magic, suddenly from the dry earth, appears a sepia-coloured knob that whitens, and ere

AGARICUS RACHODES
SHOWING GILLS AND
ANNULUS

a few days have gone has thrust its snow-white mass up like an opening parasol, edged with soft laced frill, its rounded dome trimmed with brown tabs of skin curled like a barrister's wig, its surface between softer than a bunny's throat; while underneath full three hundred clean-cut knife-edged radial gills spread around its smooth white stalk, ringed midway with its frilled annulus. And this, all this intricate beauty and more than I can tell, has sprung as it were from nothing, just while we count the hours—a wizard's whim, a conjuring trick, a miracle—revealing a

158

fancy-world, where fairies weave and dwell, and flit to rest beneath the tented shade of toadstool forests.

A TOADSTOOL FOREST

159

HARVEST AND COLOUR

NATURE paints few scenes that can compare with a tract of corn-growing country, even though it is probably inclined to flatness ; should it be undulating and, in addition, freely or only sparsely timbered, its beauty is very greatly enhanced.

Then as harvest days approach, there is an inexpressible charm in the great waving masses of upright wheat, with its brown-red-gold colouring, backed by the deep late summer green of trees against a blue sky.

THE OLD WAY—REAPING BY HAND

THE REAPERS

B.—11

DAYS IN MY GARDEN

There is a fascination in the rattling clatter of the marvellous devouring reaper as the tumbling waves of corn, like the boiling waters of a torrent, fall into its maze of mechanism, so soon to take their place in neat rows of stooks.

'THE REAPER' IN A SEA OF OATS

Walk beside the stretch of pale buff-gold oats, and see the elegant shapes and curves of its expanded glumes, the poise of the graceful nodding heads, as they rise and sway on tapered stems, pink-flushed at the base and washed with a purple bloom like that

162

THE SUNLIT GOLDEN GRAIN TAKES ITS PLACE IN NEAT ROWS OF STOOKS

11 -- 2

of a ripening plum. Or note the symmetrical rows of barley corns, in a myriad graceful arching heads, each barbed grain armed with its sharp sword-like awn, the perfect workmanship and finish of a single plant of any of these cereals : the polished tapering of the strong hollow nodose stem, the exquisite curves, the elegance and colouring designed in un-faltering beauty, and its power to extract from the earth the abundance of mar-vellous life-endowed grains ; each with its storehouse of food, containing all that is best for the sustenance of man.

In no remote hidden corner of Nature's workshop can ugliness or failure be found ; test her with the highest power of magni-fication a lens can give, there is no fault. Up on the mountain peak the great granite boulders pierce the floating clouds and seem to know the secrets of the blue beyond, the

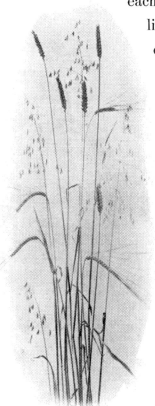

WHEAT, BARLEY, AND OATS

Drear deserts now of broken rocks,
Where Earth's bared skeleton appears,
Shattered by Frost's fierce hammer shocks,
Riven by mallets of the years !

Behold the tiny flower rooted in the narrow crevice,
see the wondrous protecting hairs that clothe its leaf,

the structure of
its pollen grains,
the sculpturing
of the seed;
Nature has not
slurred her task
nor cut her pat-
terns wrongly;
look where you
will o'er all the
earth each niche
is filled and fitted,
the fitting like
the skin upon the
hand.

CUTTING 'A ROAD' FOR THE MACHINE

Delve from her bosom store of inorganic wealth,
and look with wonder on the beauty of the crystals,
of her metals and her salts, their sparkling colours,
endowed with properties (when and why we cannot
tell) to serve the use of man.

Still there is no flaw, beauty of form is everywhere, variety of conception abounds. We are powerless to grasp the intricate greatness of the Web of Design, dumbfounded when we try to contemplate the Spinner!

THE DEEP-SEA

Read what has been found in the grappled catch from the deepest waters of the sea, where below the sparkling dancing waves lies the unthinkable region of the Deep-Sea, a place of impenetrable darkness, absolute cold stillness, imperturbable calm; yet in that eternal abode of silence and blackness, where no eye can ever catch a beam of light, there are wondrous colours, red, orange, yellow, shades of marvellous blues, but for what purpose who can say?

COLOUR

Surely Nature's colour-box is her most priceless possession, yet we are sometimes almost made to think she paints alone for bees and butterflies, and an army of creeping, crawling creatures; we lose much when we become ultra-scientific, the poetry has gone, stripped of its robe the naked facts lie gaunt and unattractive. I always like to think that with the gift of sight to man came the gift of colour also to satisfy it, 'tis too good for bees, which see not the gold of harvest fields, heed not the greens of spring-dressed woods, backed by the heavenly magnificence of a setting sun in floods of crimson fire, ere it dips below land's purple rim; they draw no honey-store from ruddy sandstone cliffs that cleave the verdant shore, where laughing blue waves, snow-crested, break in green and opal.

All thanks to the bees which have helped to heighten the brilliance of our Floral Feast, and have increased the fragrance of perfume; a noble work indeed, to which they add their gift of sweetness, and though we gain through them some faint insight of Nature's methods, we are still left to wonder why the violet flowers are violet-blue and why they are sweet, nay, when we look deeper, why they have flowers at all.

Ruskin says: 'Of all God's gifts to the sight of man, colour is the holiest and the most divine, the most solemn … and the purest and most thoughtful minds are those which love colour the most.' Everywhere in the vast world Nature's brush is at work, and when upon the Earth's face the sunbeams fall, laden with their invisible paints, we watch the colours of her canvas come and go; 'tis only when we try to copy her and imitate her matchless art that we realise her never-erring taste, and marvel at her skill; she knows no clash of colour, till man attempts the mixing.

On every side we are surrounded by this rich abundance of colour-beauty. The inimitable iridescence of the butterfly's wing, and the humming-bird's throat, the transcendent beauty of sunrise or sunset, the fired golds of autumn, the hidden, coloured world beneath the sea and earth, the limitless colour range of flowerland, flushed with every delicate shade and brilliant hue, the aforetime lights of threatening 'tempest,' the bird's egg, the opal's gleam, a thousand tones everywhere exquisitely perfect, for what purpose are they?

There are those who would label each with some utilitarian object—the square peg is ingeniously made to fit into the round hole when an awkward fact challenges a pet theory—but rather, it would seem that we see in all these the expression of the Mind of infinite beauty and grace at the back of all.

168

COLOUR

Yet we in gardens rack our brains and study many books to make our colour scheme a thing for Nature to look upon and learn from.

My companion was a young artist, whose knowledge and memory of flowers I shall ever covet. In the huge garden in which we wandered amidst its miles of trees and flowers, we were led by the controlling hand which for many years had wooed them into all their existing beauty, and my friend, realising the master mind and long experience, bombarded him with endless visionary colour schemes, until seeming almost weary the veteran exclaimed: 'Ah! these colour schemes, I have tried many—but they seldom come off—something fails.' Then pointing to a bank, crowned by a fringe of wild cherry trees, he related how the previous autumn Nature had early turned their leaves to flaming reds and golds and produced with an undergrowth—a tangled mass of michaelmas daisies —a colour scheme beyond description ; an accident.

Nature will not be ordered; if we attempt to drive her she is perverse. The varieties selected by us prove too late or too early for that particular position, flowering times do not coincide, while she from an

169

unexpected source, often with discarded and unnoticed materials, in a haphazard, almost reckless way,
weaves a scene of loveliness, flawless in its harmony
of colouring.

Yet too much consideration cannot be given to
planting with a view to future colour effect, but alas!
what failures there would often be if Nature did not
hasten in and fill the gaps, covering up our crude
efforts.

We decorate our homes with scrupulous care, the
scheme of colouring is soft, the tints subdued. Can
we introduce a bright colour without a clash? No, the
whole effect is gone. Yet we cast our eyes along the
summer border, Nature is not so careful. There the
giant spikes of Delphinium dark and light, plum
colour and smoky-purple, stand beside the fluffy,
lemon-cloudy heads of Thalictrum and flaming orange
lilies; the Madonna lily is more white, untidy Anchusa
a deeper blue, against the upright, old, scarlet Lychnis.
Tall foxgloves marked with spotted throats, and
Canterbury bells pink, white and blue, sweet-williams
and lupins—yellow lupins, tiered spikes, yellow in a
dozen shades!

COLOUR

A 'Blue Border' has a charming sound, but real blues are rare, they shade to purple, mauves, to slaty-grey-blue-whites; glaucous and silvery foliage is very lovely, but the combination lacks vigour and is cheerless.

But did I say yellow, tabooed yellow? I love its gorgeous shades, it is almost always pure, scarcely any yellow flowers show other shades in their yellow parts, and Nature gives us more wild yellow flowers at home than of any other colour. She spreads a wondrous sequence of her cloths of gold the seasons through, responsive to the first touch of earliest spring and never failing till autumn ends.

The coltsfoot, celandine and royal kingcups lead the gilded pageant in. ere spring has scarce donned her mantle of primroses, daffodils, cowslips and dandelions. Orange bird's-foot trefoil spreads a carpet beneath the gorse and broom bank's blaze; and, best of all, miles of burnished buttercups. Glorious golden days of resplendent beauty! Hawkweeds tall and short, bright corn marigolds, and ragworts reign supreme on seashore waste; there is no end of Nature's yellow throng, and no flowering tree more beautiful and bewitching than 'golden chain.'

171

TREES

I BELIEVE that our love of trees is a deeper and later development than our love of flowers, we are more mature in years when their impressive grandeur is borne in upon us. The tiny child needs no persuasion or education to love flowers: its unbounded delight in them is perfectly natural; even the hungry slum urchin, ragged and half-fed, will grab and tussle for the discarded blooms that are consigned to the dust bin, cherishing their crumpled beauty—a trait in child-life that has a deep and sacred significance.

It is only when the hand of time has silvered our locks and mellowed, but not dulled, our admiration, that there springs up within us an attachment to the trees we know. Some, maybe, that were young when we were young, others that were old, very old then, and still stand the same, breasting many a boisterous storm—we have seen them leaf and bud, cast their coats a score and more of times, and learned to know them and their ways; learned to

172

love their great, strong boles and limbs, the many-
coloured sequence of their robes, the beauty of their
quiet lives.

Somehow as we sit beneath their sheltering shade
when sorrow comes, they seem to rest and hearten
us; when we smile, are ready to dance their laughing
boughs and leaves, that dip and curtsy in the
summer breeze; have much to tell us when we stand
and listen to the music of winter's chords of many
notes sounded by their leafless fretted twigs: they
speak to us in unformed words, would almost seem to
sympathise and humour us, and so to win our hearts.

And though in flowerless days we fully appreciate the evergreen, the stiff neat fir, the polished
leaf, the wondrous shades and silvered greens of
conifers and choice shrubs, rich toned and lustrous,
it is surely the great wild trees of the tangled woods
and fields that are the most human and friendly.

Poets and writers in their adoration of flowers
have given their imagination full play, and enriched
us with a flower language, endowed them with a
power and influence, in which to speak to us; but
strong and entwined as our love may become for our

THE SYMPATHY
OF TREES

173

favourites, or for flowers as a whole, I often feel that trees are more companionable and exercise a greater sway over us: they are more individual. For though there is doubtless a difference in each dandelion or buttercup, we must look for it, while we have no difficulty in recognising many differences in each oak or beech, and the fact that many of them remind us of people we know makes them seem more human.

THE LIKENESS OF TREES
TO HUMAN BEINGS

We love the great, bold, stout-hearted oak, genuine and strong all through, as we love the same qualities in man, while there is always something in the elegant grace and rounded limbs of the beech that is womanly, noble, dignified and refined. The clean, white freshness of the young birch, with its pretty airy branches and delicate refinement— the maidens of the forest, the real girlhood of the land. The rugged stem and spreading boughs of the giant elm, storm-snapped and brittle; some men— fine men—cannot stand the sudden storm of strain and fall before it. The soft subtle silver-rippled willow, that twists and bends, to rise again—she is attractive, beautiful, but not to be trusted !

The clean hard lithe ash, that bends but will not

174

THE CLEAN WHITE FRESHNESS OF THE YOUNG BIRCH

break, strong and supple as an athlete, that with age toughens like the warrior: the stately horse-chestnut, candelabra bearer at earth's spring Offering; the stiff, straight fir, each bough in place, rigid and unbending,

.... STORM-BEATEN, BATTERED, SCARRED......

176

spick and span; the gloomy yew, the trembling aspen, and many more, how human they are!

There are some men with tough, rugged natures (fanatics, some would call them), men obsessed with one idea, filled with a noble rage to fight for some

...... STORM-TOSSED, WIND-SWEPT TREES WITH TORTURED LIMBS ......

cause they deem essential ; they stand storm-beaten, battered, scarred ; men ahead of their time, with minds advanced beyond the common measure—they are the Scotch pines of tree-land ; men whose flaming torch of individualism burns with fierceness from the dark of their mysterious lives, as when the setting sun flashes fire on the ruddy boles of pines beneath their black-plumed crowns.

SCOTCH
PINES

Often standing alone, or in a small group on some knoll or hill-top, these battered, storm-tossed, wind-swept trees with tortured limbs possess a rugged and picturesque beauty, like the misunderstood visionaries who grow old amid the turbulent storms they have raised, and possess, in their scarred but unbeaten natures, the beauty of courage undaunted, and endurance that has only been deepened by the head winds of adversity.

I never watch the sunset rays flame-flush the Scotch fir boles, their glinting lights but dimly green their massive darkened heads, distorted into beauty, but I think of those brave and noble men, who with breasts afire have faced and met a life of storm, for that which they deemed was right and true, who

178

fought, often it seemed but to fail; and passed away, to learn, perhaps, that they had conquered after all.

......THE SUNSET RAYS
FLAME-FLUSH THE SCOTCH FIR BOLES......

MIGRATION OF SWALLOWS

Days in early September are here. Blustering wind-storms and rain roughly disturb the lingering sleep of summer, and make it impossible for us to forget that 'the fall' is coming, if not upon us; storms that seem to awaken and arouse the swallows and martins, making them restless and dissatisfied as they gather in great whirling flocks and, flying high into the air, circle in an irregular manner uttering short twittering cries. Their flight is a disturbed and forced effort, a subconscious restlessness possesses them, they have little or no intention of hawking, and gradually the whole interweaving flock drifts southwards, and though the sun bursts out and summer still loiters as if loath to leave us, in a few days we realise that those which remain are but a poor remnant; most have gone from the inland districts and are gathering at their selected coast stations, for the great spirit of migration has impelled them and borne them away from us. Stream-

180

ing out somewhither they leave behind what they unconsciously dread—cold, starvation, death—and seek a land of summer and safety.

Oh wondrous expression of the power of the All-Mind that pervades every living thing! Picture the low straight flight of these birds, piercing the brine-sprayed air over the trackless sea, as they travel swiftly onward through the darkness of the night, with no landmark to guide them, meeting perhaps a sudden adverse gale of wind, yet controlled forward through all by a marvellous sense of direction, pressing on until they arrive at their haven of rest. Truly it has been said—' The appearance of a migration of shore-birds flying out of sight of land over the surface of the ocean, is indeed significant of the infinite possibilities of nature, and ideally suggests "the Spirit of God moving upon the face of the waters...." '

EARLY NOVEMBER DAYS

THERE are days in November, the month of damp,
drizzle and darkness, surpassed by few in the whole
year, days when slumbering dawn hesitates in the
heavy curtains of mist, which slowly drift and melt
away and leave a dew-drenched earth in sunshine.

THE MORNING
FOG LIFTS

From hill-top crest, already kissed by brilliant
sun, I watched a sea of mist (as if the snow-white
fleecy monsters of the heavens had fallen low and
settled on the earth), watched the shrouding, silent
mass grow restless ;—the swell of heaving soundless
waves arise and fall, then, lifting in the sunshine,
thin out to filmy wisps, which seem to struggle
and thrust out arms in vain, only to succumb, then
vanish and depart, leaving behind the sun-bathed
vales, where their soft billowy waves have been too
gentle to disturb the glories of many yet unfallen
leaves.

Lingering autumn in its marvellous later beauty
is dressed in shades of russet-copper and rusty-brown,
and in the stillness, Nature seems to stop and drink

182

in strength from the sun which now blazes in ever-conquering triumph out of the blue; a myriad of bronze-domed oaks still tightly hold their withered leaves, and hills, all bracken-patched, are bathed in glory.

The calm and silence of such days is intensified by the muteness of bird and insect life and, but for the noisy cock pheasant and occasional scolding jay, the woods are silent; their now almost leafless brushwood, provided with its winter resting-buds, fears neither wind-storm nor blizzard. In blazing attractiveness it advertises its stores of red-berried bird-food, where the untrimmed hawthorns lift their crimson blood, and scarlet hips beckon, while like patches of fire are the brittle stems of the spindle tree, with their quaintly shaped coral pink fruit disclosing its heart of gold. These are days when the sun reveals colours hidden before, days whose very short-ness and rarity seem to make them more beautiful and precious, giving, as it were, a stolen sweetness from the stormy darkness of the coming winter.

NOVEMBER

DAYS shorter still, when frost is in the air but when real winter still hesitates, and a ramble in the woods (even if not in quest of rocketing pheasant) has many charms and thrills. Heavy rain followed by frost has stripped the trees and undergrowth of their remaining leaves, and, beneath their naked beauty the damp brown leaves lie thickly on the ground, half covering banks of greenest moss, laced cushions of 'ferny' moss, and acorn cups. The thickened, lengthened twigs are not alone the leaf's past task; in falling from the hazel boughs they reveal the Promise of Spring, a host of baby catkins hang aloft, the little tight grey rolls already half-way grown to lamb's-tails. The sallow buds are fat and plump and the dark-green spurge laurel is almost in bloom. The bull-finch whistles softly to his mate and a school of long-tailed tits are searching the brushwood—their hurried stream-like flight would hardly seem to give them time for more than quaint gymnastic exercises.

HAZEL
CATKINS

THE PROMISE OF SPRING

184

NOVEMBER

Puffs of chilly air foretell the coming frost and quiet settles on the wood, as the orange-red disc of

THE RIDE IN THE WOOD

the southward-setting sun sinks in a bank of misty grey. Suddenly there is an alarm! the quick pattering

run through the fallen leaves, and the sharply uttered
'cock, cock,' accompanied by the whirr of wings,
betrays a startled old cock pheasant; the harsh
noisy cries of screaming scolding jays; a bunny
charging down the path, scenting danger, crouches a
moment, stamps and is gone; the rattling chatter of
frightened blackbird from under holly-bush, again the
wily pheasant's cry—all betoken danger; we stand
and wait: across the ride the culprit steals—Reynard,
in all his red furry magnificence, with grey-white
throat and black-tipped prick ears. 'Cock, cock':
he slinks away as if ashamed, and well he may be, for
his errands are evil and his methods are sly.

NOVEMBER
EVENING FROST

Outside the wood, we seem to feel each minute
colder than the last and hurrying darkness falls.
Once as we passed a bank of tangled broom, the
heavy crop of black seed pods sent back our thoughts
to sunny days and golden bloom, for, pausing, we
heard a running fire of little 'snicks,' as the con-
tracting cold shrank and twisted the hard seed cases.

GORSE

Back, too, went our thoughts to downs of 'blossomed
furze' when, resting

> Beside yon straggling fence that skirts the way
> With blossomed furze unprofitably gay,

186

we heard the sharp report of bursting pod, and the rattle of the falling seed. Nay, Goldsmith, thou wert wrong—'unprofitably gay'—'tis never so. Whene'er we see that sight we link our hearts with great Linnæus of old; like him, we fall upon our knees and thank God that we behold such enrapturing beauty. Such memories as these! how they cheer cold, bleak November and warm our homeward path.

But frosts on shortening days are fickle: a late and lazy dawn seems unable to penetrate the rain-burdened clouds, now so low that they mingle with the driving mists; all day long the downpour patters, there seems no dry spot to be found while Nature takes her bath; no sheltering leaves nor screening boughs, only rain, with a penetrating wetness unknown in summer days, slaking the thirst of hidden, deep-set roots, drenching the shadowed nooks and crannies, trickling into the crevices where moss and lichen quickly respond, adding charm to winter days, when the level sunlight glints through the grove and reveals a mossy-land of silvered greys passing to emerald green. Dull, damp, dark days they are indeed, when the mercurial enthusiasm of the gardener

WET DAYS

touches the minimum. The few belated flowers which have escaped the frost, present a forlorn bedraggled appearance, as they droop and mourn over their departed fellows; these battered dying remnants give no hope—they belong to a glory past, inspiring us with none of the feelings which come with the first forerunners of the early spring.

THE DEATH OF
THE FLOWERS

The melancholy days are come, the saddest of the year,
Of wailing winds, and naked woods, and meadows brown and sere.
Heaped in the hollows of the grove, the autumn leaves lie dead;
They rustle to the eddying gust, and to the rabbit's tread.

W. C. BRYANT.

WINTER DAYS

HOAR
FROST

BUT dreary days and darkest nights have their
dawn, Nature's jewelled dawn, when the rising sun
makes countless crystals flash their diamond eyes, and
clear, crisp frost holds the earth bound in silent grip.
With what marvellous skill it takes the shapeless fog
and fines it down to frosted fringe, and misses not in
all the hoary world a single bent or blade, no topmost
twig nor knotted root. In yonder copse sapling oaks
still bear some nut-brown leaves, but their heavy
burden of crystals is too great, and in the calm still
air they loose their grasp and drop in rustling flight
to earth. Alas! the sun in revealing the full beauty
of this frosted fairy-land melts its spun silver
splendour and—'tis gone!

There are two Books from whence I collect my Divinity;
besides that written one of God, another of His servant Nature,
that universal and publick Manuscript, that lies expans'd unto
the Eyes of all : those that never saw Him in the one, have
discover'd Him in the other.

SIR THOMAS BROWNE.

189

A BIT OF OLD BLUE

I WAS digging, and as each spadeful of the rich brown earth came up, I could not help wishing that the seemingly dead, yet life-giving, soil had itself feeling and life, that it might tell the secrets it withheld; big, visible life, other than that of crawling worms and countless hosts of germs. I wanted it to have life that could enjoy being brought up from the damp, smothering darkness into the glorious light and sunny air, to be reborn; then, I felt I should be like a creator, a life-giver, and for a brief space know a little of the unknown and leave the claiming world behind—but the bosom of the earth was mute.

'ALL MY HURTS
MY GARDEN SPADE CAN HEAL'

A deeper thrust than usual brought up a crumbling mass, and there below me lay a piece of finest china, blue and white, the handle on a fragment of the rim of what I knew had been long ago a rare and valued tea-cup, and thus and there by this tiny scrap, the dumb soil seemed to speak.

190

A BIT OF OLD BLUE

I saw a room, panelled and old, and through the open window grey stone steps led down to lawns, where China roses bloomed and sunbeams seemed to rest and droning bees sought the lavender bushes.

On the bare oak table silver gleamed; three cups were there, touched by the hands of three lives now long passed. In one, life's thread was much outrun, in the others, the threads were twisting into one, and she, with the soft white hands and whiter hair, smiled as she watched those strands entwine, smiled on the strong young man who knew no fear of aught in all the world, save that she whom he loved should frown on him—she filled his world, slew his courage; he would give much to be that cup touched by her lips, his all to win the hand that toyed with the handle—now but a broken scrap. I seemed to see it all as if I had been a silent watcher of the long past scene—the cup one day lay in fragments—but not the dream, for I knew this bit of blue had come to tell me so: deep love is not like china clay, but lives for ever on, unbroken, somewhere, in safe keeping.

191

THE END OF THE YEAR

ALACK! alack! How soon the year is past, the days gone, the hours all too quickly flown—except when pain and sorrow break their speed and bid them tarry in unwelcome slowness.

The folding dusk of evening curtains the sky and the spirit of the night rides in on great clouds of darkness, and real darkness enfolds 'my garden.' I stir the logs upon the hearth into a blaze, not that I need its light (for I shut my eyes that I may see more clearly), but, somehow, that ruddy flickering glow has a subtle influence which kindles memory, and before my inner eyes float the pictures of 'my garden.'

I see a river bed, grey-boulder strewn, its sun-shrunk stream a thread of whitened foam, that leaps and hides in the wadi : on every side a bare, treeless land, in colour ruddy chestnut-brown. Grouped by the stream in patchy upright growth are oleander

192

bushes, their deep rich green crowned with pink-fleshed blooms of exquisite colour, while sentinelled around upon great stones, in frequent spots as if petrified, stand quaint yet graceful herons, the lovely greys of their plumage completing one of wild Nature's perfect colour schemes.

Again, a terraced old world rose garden seems to sleep in the long June-day sun. I see and smell the fragrant blooms, I hear the happy twitter of the swallows, which like blue steel shuttles weave a great aerial web in their flight down to where the river glides.

I look down upon a flower-border into the familiar eyes of great pansies, with their cat-like faces, blending the richest hues in their soft velvet folds.

Before me lies an Irish bog, rough heather-grown, its naked patches of peat tan and sepia-toned, its face pitted where the pot-holes lurk, their margins fringed deep red with tiny sundews ; and just beyond a breeze-tossed snow-flaked breadth of cotton grass ;

the emerald greens of water grasses lace the level
stretch to where the brooms blaze and giant gorse
gold-spatters the Ben, blue-hazed. I hear the curlews
whistle, I smell the burning turves.

Anon, a silent world, snow-thatched, the burden
of deep whiteness almost too much for bending bough
and prostrate growth; there is the hush of smothered
life in the profound stillness.

I see too the familiar figure of a true friend, as
he stoops to gather carefully what some would call a
weed from the hedge-row, or, kneeling, tenderly plucks
the tiny spray from the flowerland he loves and in
which, to him, there hardly ever is a stranger. Great
sorrows have but made his life one of unbroken
praise and worship, its quiet beauty and reverence
an inspiration and an influence to those who are
privileged to know him.

And so, without any conscious mental effort, a
hundred scenes, at home, afar, visualise; each por-
trayed on the magic canvas hung in the hidden

galleries of remembrance, that no obliterating blackness can hide nor cloud dull—stored happiness; joys
that none can snatch away—memories of

Days in my garden;

mine, 'to have and to hold till death us do part' and
then still not lost: in the Larger Progress of the After
of Life to enlarge and develop more abundantly in
the Greater Gardens, where the spirit of Beauty and
of Truth shall guide and reveal to us in due time all
things that we would know.

THE END

Printed in the United States
By Bookmasters